AF825757

Dynamic Systems on Measure Chains

Mathematics and Its Applications

Managing Editor:

M. HAZEWINKEL

Centre for Mathematics and Computer Science, Amsterdam, The Netherlands

Volume 370

Dynamic Systems on Measure Chains

by

V. Lakshmikantham

Florida Institute of Technology,
Division of Interdisciplinary Programs,
Melbourne, Florida, U.S.A.

S. Sivasundaram

Embry-Riddle Aeronautical University,
Department of Mathematics,
Daytona Beach, Florida, U.S.A.

and

B. Kaymakcalan

Middle East Technical University,
Ankara, Turkey

KLUWER ACADEMIC PUBLISHERS

DORDRECHT / BOSTON / LONDON

Library of Congress Cataloging-in-Publication Data

Lakshmikantham, V.
Dynamic systems on measure chains / by V. Lakshmikantham, S. Sivasundaram, B. Kaymakcalan.
p. cm. -- (Mathematics and its applications ; v. 370)
Includes bibliographical references and index.
ISBN 0-7923-4116-3 (acid-free paper)
1. Differentiable dynamical systems. 2. Difference equations. I. Sivasundaram, S. II. Kaymakcalan, B. III. Title. IV. Series: Mathematics and its applications (Kluwer Academic Publishers) ; v. 370.
QA614.8.L34 1996
515'.352--dc20 96-9016

ISBN 0-7923-4116-3

Published by Kluwer Academic Publishers,
P.O. Box 17, 3300 AA Dordrecht, The Netherlands.

Kluwer Academic Publishers incorporates
the publishing programmes of
D. Reidel, Martinus Nijhoff, Dr W. Junk and MTP Press.

Sold and distributed in the U.S.A. and Canada
by Kluwer Academic Publishers,
101 Philip Drive, Norwell, MA 02061, U.S.A.

In all other countries, sold and distributed
by Kluwer Academic Publishers Group,
P.O. Box 322, 3300 AH Dordrecht, The Netherlands.

Printed on acid-free paper

All Rights Reserved
© 1996 Kluwer Academic Publishers
No part of the material protected by this copyright notice may be reproduced or utilized in any form or by any means, electronic or mechanical, including photocopying, recording or by any information storage and retrieval system, without written permission from the copyright owner.

Printed in the Netherlands

TABLE OF CONTENTS

PREFACE

The mathematical modeling of several important dynamic processes has been via difference equations or differential equations. Difference equations also appear in the study of discretization methods for differential equations. In recent years, however, the investigation of the theory of difference equations (discrete time dynamic systems) has assumed a greater importance as a well deserved discipline. In spite of this tendency of independence, there is a striking similarity or even duality between the theories of continuous and discrete dynamic systems. Many results in the theory of difference equations have been obtained as more or less natural discrete analogs of corresponding results of differential equations. Nevertheless, the theory of difference equations is a lot richer than the corresponding theory of differential equations. For example, a simple difference equation resulting from a first order differential equation exhibits the chaotic behavior which can only happen for higher order differential equations. Moreover, additional assumptions are often required in the discrete case in order to overcome the topological deficiency of lacking connectedness.

From a modeling point of view, it is perhaps more realistic to model a phenomenon by a dynamic system which incorporates both continuous and discrete times, namely, time as an arbitrary closed set of reals called time-scale or measure chain. It is therefore natural to ask whether it is possible to provide a framework which permits us to handle both dynamic systems simultaneously so that one can get some insight and a better understanding of the subtle differences of these two different systems. The answer is affirmative and the recently developed theory of dynamic systems on time scales offers the desired unified approach.

In this monograph, we present the current state of development of the theory of dynamic systems on time scales from a qualitative point of view. It consists of four chapters. Chapter one develops, systematically, the necessary calculus of functions on time scales following Aulbach and Hilger. In chapter two, we introduce dynamic systems on time scales and prove the basic properties of solutions of such dynamic systems. The theory of Lyapunov stability is discussed in chapter three in an appropriate setup. Chapter four is devoted to describing several different areas of investigations of dynamic systems on time scales which still await further progress. We do hope that the timely recognition of the theory of dynamic systems on time scales will provide an exciting prospect and impetus for further advances in this important area which is very new.

Some important features of the monograph are as follows:

(1) It is the first book that is dedicated to a systematic development of the theory of dynamic systems on time scales which is of recent origin;

(2) It demonstrates the interplay of the two different theories, namely, the theory of continuous and discrete dynamic systems, when imbedded in one unified framework;

(3) It provides an impetus to investigate other important problems in the setup of time scales which might offer a better understanding of the intricacies of a unified study.

In [51], the authors present a new method to control the chaotic behavior based on the evolution of a mixed logistic equation where the time flow is neither discrete nor continuous. The problem considered in [51] is clearly an example of a dynamic system on time scales with a particular time scale. This study shows that one can utilize the framework presented in this monograph to investigate general hybrid systems.

We are thankful to Professors S. Leela and Zahia Drici for helpful comments which are responsible for substantial refinement of some existing results. We wish to express our thanks to Ms. Donn Harnish for her excellent typing of this manuscript.

V. Lakshmikantham

S. Sivasundaram

B. Kaymakcalan

CHAPTER 1

1.0 Introduction

In recent times, theory of difference equations has assumed a greater importance as a well deserved discipline because of several reasons. Although there is a striking duality between the theories of differential equations and difference equations, the theory of difference equations is a lot richer than the corresponding theory of differential equations, and often demands additional assumptions to overcome the topological deficiency of lacking connectedness. None the less, it is natural to seek a framework which permits us to handle both dynamic systems simultaneously in order to get better insight and understanding of the subtle differences of the two systems. Recently developed theory of dynamic systems on time scales (closed sets of reals) provides the desired unified approach .

To meet the requirements, we shall first develop, in this chapter, systematically the calculus on time scales and utilize the developed framework in later chapters, to investigate the fundamental theory of dynamic systems on time scales as well as some qualitative properties of solutions. Thus the contents of this chapter form the core of the rest of the chapters.

We begin, in Section 1.1, with the axiomatic development of measure chains starting with any set $\mathbb{T}$ in R. We define jump operators and order topology. To remove the handicap generated by not having connectedness jump operators are necessary. We then prove an important result which includes an induction principle that is crucial for further development of calculus.

Section 1.2 introduces the notion of differentiation, and proves necessary properties of the derivatives on time scales. In Section 1.3, we discuss the mean value theorem and certain results concerning the interchange of limits and differentiation. Finally, in Section 1.4, we define suitable notion of continuity of functions, called rd-continuity, introduce the concept of anti-derivatives for rd-continuous functions, define the Cauchy-integral, and sketch different properties of the integral.

1.1 Measure chains and time scales

By a time scale we mean any closed subset of R and we denote the time scales by $\mathbb{T}$ throughout. Examples of time scales are N, Z, R, cantor set and $A = [\ t \in R : t \leq 0] \cup [\frac{1}{n} : n \in N]$. The sets Q and $B = [\ t \in R \backslash Q : 0 \leq t \leq 1]$ are not time scales.

Our aim is to develop the calculus on time scales which meets our requirements. We shall start with any set $\mathbb{T}$ and sketch the the axiomatic development. Assume that the set $\mathbb{T}$ satisfies the following axioms.

Axiom 1: *There exists a relation* $\leq$ *on* $\mathbb{T}$, *which is*

(i) *reflexive* ($t \leq t$ *for all* $t \in \mathbb{T}$),

(ii) *transitive* ($r \leq s$ *and* $s \leq t \Rightarrow r \leq t$ *for all* $r, s, t \in \mathbb{T}$),

(iii) *antisymmetric* ($r \leq s$ *and* $s \leq r \Rightarrow r = s$ *for all* $r, s \in \mathbb{T}$),

and (iv) *total* ($r \leq s$ *or* $s \leq r$ *for all* $r, s \in \mathbb{T}$).

The relation $\leq$ makes $\mathbb{T}$ a linearly ordered set or chain. we have therefore available, in chains, the notions such as boundedness, intervals, maximal elements, least upper bounds (l.u.b), greatest lower bounds (g.l.b) and order preserving mappings etc. Also, by an interval we always mean the intersection of a real interval with a given time scale and consequently, the same interval symbol can describe different time scales.

Axiom 2: *Any nonvoid subset of* $\mathbb{T}$ *which is bounded above has a least upper bound in* $\mathbb{T}$, *that is the chain* $(\mathbb{T}, \leq)$ *is conditionally complete.*

R and hZ, $h > 0$ are conditionally complete chains, while Q is not. Sometimes, we also use the measure chain $P = \bigcup_{k \in Z} [k, k + \frac{1}{2}]$.

Order Topology: The set of open intervals

$$(r, s) = [t \in \mathbb{T} : r < t < s, \;\; r, s \in \mathbb{T} \cup \{\pm\infty\}]$$

generate a topology on $\mathbb{T}$ and is called order topology on $\mathbb{T}$ with which $\mathbb{T}$ shall always be equipped. It is easy to see that a subset of $\mathbb{T}$ is compact iff it is bounded and closed and thus $\mathbb{T}$ is a locally compact topological space.

The order topology of R is well known. On hZ or N, the order topology is given by the discrete topology.

The handicap that time scales are not necessarily connected is eliminated by utilizing the notion of jump operators which we shall next define.

The mappings $\sigma, \rho : \mathbb{T}(R) \to \mathbb{T}$ defined by

$$\sigma(t) = inf[s \in \mathbb{T} : s > t], \;\; \rho(t) = sup[s \in \mathbb{T} : s < t]$$

are called the jump operators. In case, $\mathbb{T}$ is bounded above, we supplement the definition by $\sigma(max\,\mathbb{T}) = max\,\mathbb{T}$ and $\rho(min\,\mathbb{T}) = min\,\mathbb{T}$, if $\mathbb{T}$ is bounded below. The operators permit us to classify the elements of $\mathbb{T}$ as follows : $t \in \mathbb{T}$ is called

right-dense if $\sigma(t) = t$,

right-scattered if $\sigma(t) > t$,

left-dense if $\rho(t) = t$,

left-scattered if $\rho(t) < t$.

In the case $\mathbb{T} = R$, $\sigma(t) = t$, and $\mathbb{T} = hZ$, $\sigma(t) = t + h$.

A basic tool which will be employed in the proofs later is contained in the following theorem as an induction principle.

Theorem 1.1.1: *Let* $(\mathbb{T}, \leq)$ *be a chain. The following properties of* $\mathbb{T}$ are *equivalent:*

(a) $\mathbb{T}$ *is conditionally complete.*

(b) *Any nonvoid subset* S *which is bounded below has a g.l.b.*

(c) *The induction principle holds:*

Assume that for a family of statements $A(t)$, $t \in [\tau, \infty) \subseteq \mathbb{T}$

the following conditions are satisfied:

(I) $A(\tau)$ *holds true.*

(II) *for each right-scattered* $t \in \mathbb{T}$ *we have* $A(t) \Rightarrow A(\sigma(t))$.

(III) *for each right-dense* $t \in \mathbb{T}$ *there is a neighborhood* U *such that* $A(t) \Rightarrow A(s)$ *for all* $s \in U$, $s > t$.

(IV) *for each left-dense* $t \in \mathbb{T}$ *one has* $A(s)$ *for all* s *with* $s < t \Rightarrow A(t)$

Then $A(t)$ *is true for all* $t \in [\tau, \infty)$.

(d) *A subset* $S \subseteq \mathbb{T}$ *is compact, if and only if it is bounded and closed.*

(e) *The intermediate value theorem holds: The continuous mapping* $f : [r, s] \to R$, $r, s \in \mathbb{T}$, *is assumed to fulfill the condition*

$$f(r) < 0 < f(s) \tag{1.1.1}$$

Then there is a $\tau \in [r, s]$ *with*

$$f(\tau).f(\sigma(\tau)) \leq 0. \tag{1.1.2}$$

Remarks: For $\mathbb{T} = N$ the conditions (III) and (IV) of the induction principle are redundant. In this case it reduces to the usual principle of mathematical induction for the natural numbers. If $\mathbb{T}$ is a left closed interval in R, then condition (II) becomes redundant. This induction principle then contains the (topological) principle that a nonvoid subset of an interval which is at the same time open and closed coincides with this interval. This continuous type of induction is used by Dieudonne [11] for proving basic results in differential and integral calculus.

Proof: (a) $\Rightarrow$ (b): Choose $inf\, S := sup\{t \in \mathbb{T} : t < s \, for\, all\, s \in S\}$.

(b) $\Rightarrow$ (c): Assume that there is some $t \in [\tau, \infty)$ such that $A(t)$ does not hold. Then we define

$$r = inf\{\, t \in [\tau, \infty) \colon A(t)\;\; does\, not\, hold\} \tag{1.1.3}$$

(1) $A(r)$ is valid, therefore $A(t)$ holds true for all $t \in [\tau, r\,]$.

Case 1: r is left-scattered. $A(\rho(r))$ holds true, with (II) it follows $A(r) = A(\sigma(\rho(r)))$

Case 2:. r is left-dense. The assertion follows by (IV).

Case 3: r is minimal. Then we have r = τ, which leads to the assertion by (I).

(2) In each of the following cases we will obtain a contradiction.

Case 1: r is right-scattered. From $A(r)$ we conclude by (II), $A(\sigma(r))$ and then $r < \sigma(r) \leq r$.

Case 2: r is right-dense. Each neighborhood of $r = \sigma(r)$ contains some interval $[r, s]$ with $s > r$. From (1) we have $A(r)$ and then with (III) $A(t)$ for all $t \in [r, s]$, even for all $t \in [\tau, s]$ by (1.1.3). It follows $r < s \leq r$.

Case 3: r is maximal. With (1) we have $A(t)$ for all $t \in [\tau, r]$.

(c) ⇒ (d): Let $[\tau, s]$ be some bounded and closed interval and $(U_i)_{i\in I}$ an arbitrary open covering for $[\tau, s]$. We show by means of the induction principle for $t \in [\tau, s]$:

$$A(t) : [\tau, t] \text{ has a finite subcovering.}$$

The verification of (I), (II), (III) are trivial. For (IV), first of all we choose some U_j from the covering with $t \in U_j$ and then some $r \in U_j$ with $r < t$. Such an r exists, because t is left-dense. The finite subcovering of $[\tau, r]$ together with U_j forms a finite subcovering for $[\tau, t]$.

Each bounded and closed subset of $\mathbb{T}$ is a closed subset of some compact interval and therefore compact itself. If not a compact S is unbounded, and one can immediately find an infinite open covering without a finite subcovering. Finally each compact subset is closed.

(d) $\Rightarrow$ (e): We assume that (1.1.2) does not hold, hence:

$$f(\tau).f(\sigma(\tau)) > 0 \text{ for all } \tau \in \mathbb{T} \tag{1.1.4}$$

The set $A := f^{-1}((-\infty, 0))$ and the set B of the upper bounds of A in $[r, s]$ are nonvoid. We choose $p \in A$, $q \in B$.

For a finite collection of closed intervals $[a_i, b_i] \subseteq [p, q]$ with $a_i \in A$, $b_i \in B$, $i \in J$, J finite, we have:

$$\bigcap_{i \in J} [a_i, b_i] = [max\{a_i\}, min\{b_i\}] \neq \phi. \tag{1.1.5}$$

Moreover, because of the compactness of $[p, q]$, $\bigcap_{a \in A, b \in B} [a, b] \neq \phi$.

Otherwise the dual to the Heine Borel property would guarantee that even a finite intersection of these intervals would be empty, contradicting (1.1.5). Therefore there is some $c \in [r, s]$ with $a \leq c \leq b$ for all $a \in A$, $b \in B$.

Each neighborhood of $\sigma(c)$ contains some t with $t > c$ or $t = s$, in any case some t with $f(t) > 0$. The continuity of f has the consequence $f(\sigma(c) \geq 0$, and then by (1.1.4) , $f(c) > 0$, c cannot be left-dense. In this case each neighborhood U of c would meet the set A and this contradicts the continuity of f in c.

If c is left-scattered, then $\rho(c)$ is in A. Then with $\tau = \rho(c)$, $\sigma(\tau) = c$ the assertion (1.1.4) cannot be claimed anymore.

(e) $\Rightarrow$ (a): Let S be some nonvoid subset of $\mathbb{T}$ which is bounded above. We define B to be the nonvoid set of upper bounds of S and A to be the nonvoid set of lower bounds of B. Choose $r \in A$, $s \in B$ with $r < s$ (w.l.o.g. $\mathbb{T}$ contains more than two elements). We define the mapping $f : [r, s] \to R$ by

$$f(t) = \begin{cases} -1, & if\ t \in A, \\ +1, & if t \in B. \end{cases}$$

If A does not posses any l.u.b., hence B any g.l.b., then the sets

$$f^{-1}(\{-1\}) = \bigcup_{t \in A} [r, t), \qquad f^{-1}(\{+1\}) = \bigcup_{t \in B} (t, s]$$

are open in $[r, s]$ and therefore f is continuous. By (e) there is some $\tau \in [r, s]$ with $f(\tau).f(\sigma(\tau)) \leq 0$. Because of the fact that f does not vanish, we get $f(\tau).f(\sigma(\tau)) < 0$ and therefore $f(\tau) < 0 < f(\sigma(\tau))$. Thus A does have a l.u.b., which is one of S, too. The proof of theorem is complete.

1.2 Differentiation

In order to create a notion of differentiation for functions that are defined on conditional complete chains, we need to measure distance between two elements of $\mathbb{T}$. The following axiom offers this facility.

Axiom 3: *On a conditionally complete chain* $\mathbb{T}$, *there exists a mapping* $\mu : \mathbb{T} \times \mathbb{T} \rightarrow R$ *which satisfies the properties*

(i) $\mu(r,s) + \mu(s,t) = \mu(r,t),\ t, r, s \in \mathbb{T}, (cocylic\ property)$

(ii) $\mu(r,s) > 0\ if\ r > s, \quad r, s \in \mathbb{T}, (strong\ isotony),$

(iii) $\mu\ is\ continuous.\ (continuity).$

In view of the property (i) , we have $\mu(r,s) = \mu(r,t) - \mu(s,t)$, for $t, r, s \in \mathbb{T}$, and hence the continuity of μ is equivalent to the fact that $\mu(.,t)$ is continuous for fixed $t \in \mathbb{T}$, One can show that μ induces a measure on $\mathbb{T}$ by assigning $\mu([r,s)) = \mu(s,r)$. We say that the triple $(\mathbb{T}, \leq, \mu)$ or the pair $(\mathbb{T}, \mu)$ is a measure chain with time scale $\mathbb{T}$ and growth calibration μ.

The growth calibration has the following consequence relative to the topology on $\mathbb{T}$.

Theorem 1.2.1: *If* $(\mathbb{T}, \mu)$ *is a measure chain, then* $\mathbb{T}$ *(equipped with the order topology) is metrizable and a* K_σ*-space, i.e. it is a union of at most denumerably many compact sets. The metric is given by:*

$$d(r,s) = \left|\mu(r,s)\right| \tag{1.2.1}$$

Proof: For a fixed $\tau \in \mathbb{T}$, set $\mu(.) = \mu(.,\tau)$.

(1) An open neighborhood (r,s) of τ contains an open ϵ – ball $U_\epsilon(\tau)$ with $\epsilon = min\{\left|\mu(r,\tau)\right|, \left|\mu(s,\tau)\right|\}$. Thus the metric topology is finer than the order topology. On the other hand, by the continuity of μ, there is always an order open inverse image of the set $(-\epsilon, +\epsilon)$ contained in an ϵ-ball $U_\epsilon(\tau)$.

(2) For $n \in N$, we define the real numbers:

$$a_n := \begin{cases} -n \\ a + \frac{1}{n} \\ a \end{cases} \quad if\ inf\,\mu(\mathbb{T}) = \begin{cases} -\infty \\ a \notin \mu(\mathbb{T}) \\ a \in \mu(\mathbb{T}) \end{cases},$$

$$b_n := \begin{cases} +n \\ b - \frac{1}{n} \\ b \end{cases} \quad if\ sup\,\mu(\mathbb{T}) = \begin{cases} +\infty \\ b \notin \mu(\mathbb{T}) \\ b \in \mu(\mathbb{T}) \end{cases},$$

Then for each $n \in N$, the set $\mu^{-1}([\, a_n, b_n \,])$ is bounded and closed, hence compact, and we have $\mathbb{T} = \bigcup_{n=1}^{\infty} \mu^{-1}([\, a_n, b_n \,])$ and the proof is complete.

Observing the K_σ -property we can form the expressions $\lim_{t \to \pm\infty} f(t) = a$ for a function f with values in a metric space by means of sequences.

For an arbitrary conditionally complete chain $\mathbb{T}$ we define

$$\mathbb{T}^k : = \{ t \in \mathbb{T} : t \ nonmaximal \ or \ left - dense \} \tag{1.2.2}$$

Thus this operator cuts off an eventually existing isolated maximum of $\mathbb{T}$. The function

$$\mu^*\colon \mathbb{T}^k \to R_0^+ \text{ defined by } \mu^*(t) = \mu\,(\sigma(t), t), \tag{1.2.3}$$

is called *graininess*. For (R) we have $\mu^* \equiv 0$, for (hZ) we have $\mu^* \equiv h$.

We call a triple $(\mathbb{T}, \mu, X)$ a dynamical triple, if $(\mathbb{T}, \mu)$ is a measure chain and X is a Banach space.

Let f be a mapping $\mathbb{T} \to X$. At $t \in \mathbb{T}$, f has the derivative $f_t^\Delta \in X$, if for each $\epsilon > 0$, there exists a neighborhood U of t such that for all $s \in U$

$$\left| f(\sigma(t)) - f(s) - f_t^\Delta \mu.(\sigma(t), s) \right| \le \epsilon \left| \mu.(\sigma(t), s) \right|. \tag{1.2.4}$$

f is called *differentiable* at $t \in \mathbb{T}$, if f has exactly one derivative f_t^Δ in t.

It is easy to check that in case (R) these notions coincide with those of ordinary differential calculus. For (hZ) each mapping $f : hZ \to X$ is differentiable in each $t \in hZ$ and we have

$$f_t^\Delta = \Delta_h f(t) = [f(t+h) - f(t)]/h, \tag{1.2.5}$$

which is the difference operator. In the literature difference equations are sometimes written in the form $u(t+1) = g(u(t))$ or $u_{n+1} = g(u_n)$. However, in the present frame work, we need to use the difference operator to represent difference equations. Hence care must be taken when deducing the results corresponding to difference equations.

The following theorem collects the basic properties of derivative.

Theorem 1.2.2: Let f be a mapping $\mathbb{T} \to X$ and $t \in \mathbb{T}$.

(i) *If* $t \in \mathbb{T}^k$, *f has at most one derivative in t.*

(ii) *If* $t \notin \mathbb{T}^k$, *f has each* $x \in X$ *as a derivative*

(iii) *If f has a derivative in t, then f is continuous in t.*

(iv) *The mapping* $\mu.(.,\tau)$ *is differentiable in t with*

$$\mu\,(.,\tau)_t^\Delta = 1. \tag{1.2.6}$$

(v) *If t is right − scattered and f is continuous in t, then f is differentiable in t.*

Then we have

$$f_t^\Delta = \frac{f(\sigma(t)) - f(t)}{\mu^*(t)}. \tag{1.2.7}$$

Proof: (i) Assume that f has two derivatives $x, y \in X$ in t. By definition of $\mathbb{T}^k$, each neighborhood of t contains some $s \in \mathbb{T}^k$, with $s \neq \sigma(t)$. Hence, for each $\epsilon > 0$, there exists some $s \neq \sigma(t)$ with

$$\left|f(\sigma(t)) - f(s) - x\,\mu.(\sigma(t), s)\right| \leq \tfrac{1}{2}\epsilon\left|\mu.(\sigma(t), s)\right| \;,$$

$$\left|f(\sigma(t)) - f(s) - y\,\mu.(\sigma(t), s)\right| \leq \tfrac{1}{2}\epsilon\left|\mu.(\sigma(t), s)\right| .$$

This implies

$$\left|x - y\right| . \left|\mu.(\sigma(t), s)\right| = \left|(x - y)\,\mu.(\sigma(t), s)\right|$$

$$\leq \left|x\,\mu(\sigma(t), s) - [f(\sigma(t)) - f(s)]\right| + \left|f(\sigma(t)) - f(s) - y\,\mu.(\sigma(t), s)\right|$$

$$\leq \epsilon\left|\mu.(\sigma(t), s)\right|.$$

Observing that $\mu.(\sigma(t), s) \neq 0$, we get $\left|x - y\right| \leq \epsilon$ and hence $x = y$.

(ii) In this case, for each $\epsilon > 0$ and for each $x \in X$ there exists a neighborhood U (namely $U = \{t\}$) of $t = \sigma(t)$ such that for all $s \in U$ (hence $s = t$) we have

$$\left|f(t) - f(s) - x\,\mu.(t, s)\right| = 0 \leq \epsilon\left|\mu.(t, s)\right|.$$

(iii) Let $t \in \mathbb{T}$ and U be a compact neighborhood of f. We define

$$M = max(\{\left|\mu.(\sigma(t), s)\right| : s \in U\} \cup \{\left|f_t^{\Delta}\right|\}) > 0 \,.$$

For $\epsilon > 0$ there exists a neighborhood V of t such that for all $s \in V$

$$\left|f(\sigma(t)) - f(s) - f_t^{\Delta}\mu.(\sigma(t),s)\right| \leq \frac{\epsilon}{4M}\left|\mu.(\sigma(t),s)\right|,$$

$$\left|\mu.(\sigma(t),s) - \mu.(\sigma(t),t)\right| \leq \frac{\epsilon}{2M}.$$

Thus for $s \in U \cap V$, $\left|f(s) - f(t)\right|$

$$= \left|f(s) - f(\sigma(t)) + f_t^{\Delta}\mu.(\sigma(t),s) + f(\sigma(t)) - f(t) - f_t^{\Delta}\mu.(\sigma(t),t)\right.$$

$$\left. + f_t^{\Delta}[\mu.(\sigma(t),t) - \mu.(\sigma(t),s)]\right| \leq \frac{\epsilon}{4M}.(\left|\mu.(\sigma(t),s)\right| + \left|\mu.(\sigma(t),t)\right|)$$

$$+ \left|f_t^{\Delta}\right|\left|\mu.(\sigma(t),t) - \mu.(\sigma(t),s)\right|$$

$$\leq \frac{\epsilon}{4M}.2M + \frac{\epsilon}{2M}.M = \epsilon.$$

(iv) Follows from property (i) of $\mu.$ in axiom 3.

(v) The mapping $g(.) := \frac{f(\sigma(t)) - f(.)}{\mu(\sigma(t),.)}$ is defined and continuous in a neighborhood of t.

For $\epsilon > 0$ there exists a neighborhood U of t such that for $s \in U$

$$\left|f(\sigma(t)) - f(s) - \frac{f(\sigma(t)) - f(t)}{\mu^*(t)}.\mu.(\sigma(t),s)\right|$$

$$= \left|\frac{f(\sigma(t)) - f(s)}{\mu(\sigma(t),s)} - \frac{f(\sigma(t)) - f(t)}{\mu(\sigma(t),t)}\right|.\left|\mu.(\sigma(t),s)\right| \leq \epsilon\left|\mu.(\sigma(t),s)\right|.$$

The uniqueness of the derivative follows from (i) and the proof is complete.

The algebraic properties, such as linearity, Leibniz's product rule, and quotient rule for the derivative are given in the next result.

Theorem 1.2.3:

(i) *Let* f *and* $g: \mathbb{T} \to X$ *be differentiable in* $t \in \mathbb{T}^k$. *Then for* $\alpha, \beta \in K$ *the mapping* $\alpha f + \beta g$ *is differentiable in* t, *and we obtain*

$$(\alpha f + \beta g)_t^{\Delta} = \alpha f_t^{\Delta} + \beta g^{\Delta}{}_t \tag{1.2.8}$$

(ii) *Let* X_1, X_2, X_3 *be Banach spaces and* $. : X_1 \times X_2 \to X_3$ *be some bilinear and continuous mapping. If* $f: \mathbb{T} \to X_1$ *and* $g: \mathbb{T} \to X_2$ *are differentiable in* $t \in \mathbb{T}^k$ *so is* $f.g$, *and we have:*

$$(f.g)_t^{\Delta} = f(\sigma(t)).g_t^{\Delta} + f_t^{\Delta}.g(t). \tag{1.2.9}$$

(iii) *Let* X *be a Banach algebra with 1. The mappings* $f, g : \mathbb{T} \to X$ *are assumed to be algebraically invertible with* $f.g = 1$. *If* f *is differentiable in* $t \in \mathbb{T}^k$. *then* g *is differentiable, too, and we get:*

$$g_t^{\Delta} = -g(\sigma(t)).f_t^{\Delta}.g(t). \tag{1.2.10}$$

(iv) *Let* $V : \mathbb{T} \times R^n \to R^n$ *and* $x : \mathbb{T} \to R^n$ *be two functions such that* x^{Δ}, V_t^{Δ} *and* V_x *exist. Then*

$$[V(t,x(t))]_t^{\Delta} = V_t^{\Delta}(t,x(t)) + V_x(\sigma(t),x(t))\; x_t^{\Delta}.$$

Proof: (i) is trivial.

(ii) Let an arbitrary $\epsilon > 0$ be given. Choose $\epsilon_1, \epsilon_2, \epsilon_3 > 0$ such that

$$|f(\sigma(t))|\epsilon_1 + (|g(t)| + \epsilon_3)\epsilon_2 + |f_t^{\Delta}|\epsilon_3 \le \epsilon$$

and some neighborhood U of t such that for all $s \in U$,

$$|g(s) - g(\sigma(t)) + g_t^{\Delta}\mu.(s,\sigma(t))| \le \epsilon_1 |\mu.(s,\sigma(t))|,$$

$$|f(s) - f(\sigma(t)) + f_t^{\Delta}\mu.(s,\sigma(t))| \le \epsilon_2 |\mu.(s,\sigma(t))|,$$

$$|g(s) - g(t)| \le \epsilon_3.$$

Hence, for $s \in U$

$$|f(s).g(s) - f(\sigma(t)).g(\sigma(t)) - [f(\sigma(t))g_t^{\Delta} + f_t^{\Delta}g(t)]\mu.(s,\sigma(t))| \le$$

$$|f(\sigma(t)).[g(s) - g(\sigma(t)) - g_t^{\Delta}\mu.(s,\sigma(t))] +$$

$$[f(s) - f(\sigma(t)) - f_t^{\Delta}\mu.(s,\sigma(t))][g(t) + g(s) - g(t)] +$$

$$f_t^{\Delta}[g(s) - g(t)]\mu.(s,\sigma(t))| \le$$

$$[|f(\sigma(t))|\epsilon_1 + (|g(t)| + \epsilon_3)\epsilon_2 + |f_t^{\Delta}|\epsilon_3]|\mu.(s,\sigma(t))| \le \epsilon|\mu.(s,\sigma(t))|.$$

(iii) The mapping $X^* \to X^*$, $X \to X^{-1}$ within the multiplicative group of X is continuous. Therefore $g(.)$ is continuous in t just as $f(.)$. For an arbitrary $\epsilon > 0$, there exist ϵ_1, ϵ_2 such that

$$|g(\sigma(t))|\{\epsilon_2(\epsilon_1 + |g(t)|) + \epsilon_1|f_t^{\Delta}|\} \quad \le \epsilon$$

and some neighborhood U of t such that for $s \in U$,

$$|g(s) - g(t)| \leq \epsilon_{1},$$

and

$$|f(s) - f(\sigma(t)) + f_t^{\triangle}\mu.(s, \sigma(t))| \leq \epsilon_2 |\mu.(s, \sigma(t))|.$$

Now we get for $s \in U$,

$$|g(s) - g(\sigma(t)) + g(\sigma(t)) f_t^{\triangle} g(t)\mu.(s, \sigma(t))|$$

$$=| - g(\sigma(t)).\{[f(s) - f(\sigma(t)) - f_t^{\triangle}\mu.(s, \sigma(t))]g(s)$$

$$+ f_t^{\triangle}[g(s) - g(t)]\mu.(s, \sigma(t))\}|$$

$$\leq |g(\sigma(t))|\{\epsilon_2(\epsilon_1 + |g(t)|) + \epsilon_1 |f_t^{\triangle}|\}.|\mu.(s, \sigma(t))|$$

$$\leq \epsilon|\mu.(s, \sigma(t))|.$$

The proof of (iv) can be constructed following the foregoing proofs and hence we omit the details. The proof of Theorem is therefore complete.

1.3 Mean Value Theorem and Consequences

The mapping $f : \mathbb{T} \to X$ is called pre-differentiable on $\mathbb{T}$ with region of differentiation D, if for the pair (f, D) the following conditions hold:

(i) $D \subseteq \mathbb{T}^k$, $\mathbb{T}^k \backslash D$ is at most denumerable and contains no right-scattered element of $\mathbb{T}$.

(ii) f is continuous on $\mathbb{T}$ and differentiable in each $t \in D$.

In other words, f is continuous and everywhere differentiable with the exception of at most denumerable many right-dense points.

The mapping f is called differentiable on $\mathbb{T}$ if it is pre-differentiable with $D = \mathbb{T}^k$. The mapping $f^\Delta : D \to X$, $t: \to f_t^\Delta$ is called pre-derivative with the range of definition D resp. derivative of f.

Example: If $\mathbb{T}$ is an arbitrary interval of R, then a function which is differentiable on $\mathbb{T}$, is differentiable in the interior of $\mathbb{T}$ and differentiable with respect to one side in end points belonging to $\mathbb{T}$.

The notion of pre-differentiability is introduced only for the proof of Theorem 1.4.4 which assures the existence of antiderivatives. we will not need it, when this theorem is proved. Nevertheless the theorems following immediately are the basis for the development of the analysis on measure chains when applying them for differentiable functions.

We are now in a position to prove the mean value theorem.

Theorem 1.3.1: *Let the mapping* $f : \mathbb{T} \to X$, $g : f : \mathbb{T} \to R$ *be pre-differentiable with* D, *and assume that*

$$|f^\Delta(t)| \le g^\Delta(t),\ t \in D. \tag{1.3.1}$$

Then for $r, s \in \mathbb{T}$, $r \leq s$:

$$|f(s) - f(r)| \leq g(s) - g(r). \tag{1.3.2}$$

Proof: There is a surjective mapping $N \to [r,s)\backslash D,\ n \to t_n$. Let $\epsilon > 0$ be given. we apply the induction principle in $[r,s]$ to the statements

$$A(t):\ |f(s) - f(r)| \leq g(s) - g(r) + \epsilon[\mu.(t,r) + \sum_{t_n<t} 2^{-n}].$$

From $A(s)$ for arbitrary $\epsilon > 0$ the claimed inequality (1.3.2) follows.

(I) The statement $A(r)$ is trivially satisfied.

(II) Let t be right-scattered, hence $t \in D$. By (1.2.7) we have

$$f(\sigma(t) - f(t) = f^{\Delta}(t).\mu^*(t) \text{ and } g(\sigma(t)) - g(t) = g^{\Delta}(t).\mu^*(t).$$

whence, observing that $|f^{\Delta}(t)| \leq g^{\Delta}(t)$,

$$|f(\sigma(t) - f(t)| \leq g(\sigma(t)) - g(t).$$

The addition of this inequality to that of the induction condition $A(t)$ yields $A(\sigma(t))$.

(III) Let t be right-dense so that $\sigma(t) = t \neq s$.

Case 1: Let $t \in D$. Then there exists some neighborhood U of t such that for all $p \in D$ the inequalities

$$\left|f(t)-f(p)+f^{\Delta}\mu.(t,p)\right| \leq \tfrac{1}{2}\epsilon\left|\mu.(t,p)\right|,$$

$$\left|g(t)-g(p)+g^{\Delta}\mu.(t,p)\right| \leq \tfrac{1}{2}\epsilon\left|\mu.(t,p)\right|,$$

hold true, whence the estimates

$$\left|f(p)-f(t)\right| \leq (\left|f^{\Delta}(t)\right|+\tfrac{1}{2}\epsilon).\left|\mu.(t,p)\right|,$$

$$g(p)-g(t)+\tfrac{1}{2}\epsilon\left|\mu.(t,p)\right| \geq g^{\Delta}(t).\mu.(p,t),$$

follow. For all $p \in U$, $p \geq t$ we get

$$\left|f(p)-f(t)\right| \leq (\left|f^{\Delta}(t)\right|+\tfrac{1}{2}\epsilon).\mu.(p,t)$$

$$\leq (g^{\Delta}(t)+\tfrac{1}{2}\epsilon).\mu.(p,t) \leq g(p)-g(t)+\epsilon.\mu.(p,t).$$

Case 2: Let $t \in [r,s)\backslash D$. There exists an $m \in N$ with $t = t_m$ and some neighborhood U of t such that for $p \in U$ we have,

$$\left|f(p)-f(t)\right| \leq \tfrac{\epsilon}{2}.2^{-m} \text{ and } \left|g(p)-g(t)\right| \leq \tfrac{\epsilon}{2}.2^{-m}.$$

From the second inequality we obtain

$$0 \leq g(p)-g(t)+\tfrac{\epsilon}{2}.2^{-m}.$$

An addition of this inequality to the first one yields

$$\left|f(p)-f(t)\right| \leq g(p)-g(t)+\epsilon.2^{-m}.$$

Therefore in both cases there exists some neighborhood U of t such that for $p \in U$, $p \geq t$ we have,

$$|f(p) - f(t)| \leq g(p) - g(t) + \epsilon(\mu.(p,t) + \sum_{t \leq t_n < p} 2^{-n}).$$

Taking into account the triangle inequality we get the statements $A(p)$ for all $p \in U$, $p \geq t$ by adding this inequality to the inequality of the induction condition $A(t)$.

(IV) Observing that the $\leq$ inequality between continuous functions is valid on $[r,t)$, we conclude that it is valid on $[r,t]$, too.

Corollary 1.3.1: (i) *Let $f : \mathbb{T} \to X$ be pre-differentiable with D, and U be some compact interval of $\mathbb{T}$ with endpoints r and s. Then we have*

$$|f(s) - f(r)| \leq \sup_{t \in U^k \cap D} \{|f^{\Delta}(t)|\}.|\mu.(s,r)|. \tag{1.3.3}$$

(ii) *If $f : \mathbb{T} \to X$ be pre-differentiable with D and $f^{\Delta}(t) = 0$ for all $t \in D$, then f is constant.*

Proof: For (i) we can assume without loss of generality, that $r \leq s$. Consider the right side as a function g of s and apply the mean value theorem. (ii) immediately follows from (i).

Concerning the interchange of limits and differentiation, we have the following result.

Theorem 1.3.2: *Let the functions* $f_n : \mathbb{T} \to X$ *of the sequence* $(f_n)_{n\in N}$ *be pre-differentiable with* D. *Assume that for each* $t \in \mathbb{T}^k$ *there exists a compact interval neighborhood* U_t *such that the sequence of the pre-derivatives* $(f_n^{\Delta})_{n\in N}$ *converges uniformly on* $U_t \cap D$. *Then*

(i) *The convergence of* $(f_n)_{n\in N}$ *in one* $t_k \in \mathbb{T}$ *implies the uniform convergence of this sequence on each* U_t.

(ii) $f := \lim_{n\to\infty} f_n$ *is pre-differentiable with* D, *and we have*

$$f^{\Delta}(t) = \lim_{n\to\infty} f_n^{\Delta}(t),\ t \in D. \tag{1.3.4}$$

Proof: We can assume that $\sigma(t) \in U_t$ for each t.

(1) Let $t \in \mathbb{T}^k$. We show that if $(f_n(\tau))_{n\in N}$ converges for one $\tau \in U_t$, then $(f_n)_{n\in N}$ converges uniformly on U_t. First of all, there exists an $n_0 \in N$ such that

$$\sup_{s \in U_t \cap D} \{|(f_n - f_m)^{\Delta}(s)|\}$$

exists for all $n, m \geq n_0$. For an arbitrary $r \in U_t$, $n, m \geq n_0$ by corollary 1.3.1 (i) we get

$$\begin{aligned}
&\left|f_n(r) - f_m(r)\right| - \left|f_n(\tau) - f_m(\tau)\right| \\
&\leq \left|[f_n(r) - f_m(r) - f_n(\tau) - f_m(\tau)]\right| \qquad (1.3.5) \\
&\leq \sup_{s \in U_t \cap D} \{\left|(f_n - f_m)^{\Delta}(s)\right|\}.\left|\mu.(\tau, r)\right| .
\end{aligned}$$

This implies

$$\begin{aligned}
&\left|f_n(r) - f_m(r)\right| \leq \left|f_n(\tau) - f_m(\tau)\right| \\
&\quad + \sup_{s \in U_t \cap D} \{\left|(f_n^{\Delta}(s) - f_m^{\Delta}(s))\right|\}.\left|\mu.(max\, U_t, min\, U_t)\right|\}.
\end{aligned}$$

The right side of this inequality can be made arbitrarily small, if n, m are large enough. The numbers n, m can be chosen independently of $r \in U_t$, thus we get the claimed assertion, since X is complete.

(2) We want to show that for each $t \in \mathbb{T}$, $(f_n(t))_{n \in N}$ converges.

For this purpose, we apply the induction principle in $[t_k, \infty)$ to the statements

$A(t)$: $(f_n(t))_{n \in N}$ converges.

(I) In fact $(f_n(t_k))_{n \in N}$ converges.

(II) Let t be right-scattered, hence $t \in D$. With (1.2.7) we see that for $n \in N$

$$f_n(\sigma(t)) = f_n(t) + f_n^{\Delta}(t).\mu^*(t).$$

By the assumption of the theorem $(f_n^{\Delta}(t))_{n \in N}$ converges. Therefore from $A(t)$ the statement $A(\sigma(t))$ follows.

(III) and (IV). These conditions can easily be checked by the statement (1).

For (IV) observe that U_t always contains some s with $s < t$, if t is left-dense. Hence, we have $A(t)$ for $t \geq t_k$.

For $t \leq t_k$ the assertion can be shown in the same way. One applies the dual version of the induction principle for the negative direction. Thereby the roles of ρ and σ change. Only when proving condition (II) one has to state equation (1.2.7) as follows

$$f_n(\rho(t)) = f_n(t) + f_n^{\Delta}(\rho(t)).\mu^*(\rho(t)).$$

(3a) f is continuous as a limit function of a locally uniformly convergent sequence of continuous functions. Let $g : D \to X$ be defined by

$$g(s) : = \lim_{n \to \infty} f_n^{\Delta}(s).$$

Now, let $t \in D$ and $\epsilon > 0$ be given. Then there is some $n_1 \geq n_0$ (Cf. (1)) such that for $n, m \geq n_1$ we get

$$\sup_{s \in U_t \cap D} \{|(f_n^{\Delta}(s) - f_m^{\Delta}(s))|\} \leq \tfrac{\epsilon}{3}$$

In inequality (1.3.5), replace τ by $\sigma(t)$, so that for all $r \in U_t$ and $n, m \geq n_1$

$$|[f_n(r) - f_m(r)] - [f_n(\sigma(t)) - f_m(\sigma(t))]|$$

$$\leq \sup_{s \in U_t \cap D} \{|(f_n^{\Delta}(s) - f_m^{\Delta}(s))|\}|\mu.(\tau, \sigma(t))| \leq \tfrac{\epsilon}{3}.|\mu.(\tau, \sigma(t))|.$$

Hence as $m \to \infty$, $n \geq n_1$, it follows

$$\left|[f_n(r) - f(r)] - [f_n(\sigma(t)) - f(\sigma(t))]\right| \leq \tfrac{\epsilon}{3}.\left|\mu.(\tau, \sigma(t))\right|. \tag{1.3.6}$$

(3b) There exists some $j \geq n_1$ such that

$$\left|[f_j^{\Delta}(\sigma(t)) - g(\sigma(t))]\right| \leq \tfrac{\epsilon}{3}. \tag{1.3.7}$$

(3c) f_j is differentiable in t, therefore we have a neighborhood W of t such that for all $r \in W$

$$\left|f_j(\sigma(t)) - f_j(r) - f_j^{\Delta}(t)\mu.(\sigma(t), r)\right| \leq \tfrac{\epsilon}{3}.\left|\mu.(\tau, \sigma(t))\right|. \tag{1.3.8}$$

(4) Now, for $r \in U_t \cap W$, it follows from (1.3.6), (1.3.7) and (1.3.8),

$$\begin{aligned}
&[f(\sigma(t)) - f(r) - g(t)\,\mu.(\sigma(t), r)| \\
&\quad \leq \left|[f(\sigma(t)) - f(r)] - [f_j(\sigma(t)) - f_j(r)]\right| \\
&\qquad + \left|f_j(\sigma(t)) - f_j(r) - f_j^{\Delta}(t)\mu.(\sigma(t), r)\right| \\
&\qquad + \left|[f_j^{\Delta}(\sigma(t)) - g(\sigma(t))].\mu.(\sigma(t), r)\right| \\
&\quad \leq (\tfrac{\epsilon}{3} + \tfrac{\epsilon}{3} + \tfrac{\epsilon}{3}).\left|\mu.(\sigma(t), r)\right| = \epsilon.\left|\mu.(\sigma(t), r)\right|.
\end{aligned}$$

The proof is therefore complete.

1.4 Integral and Antiderivative

Let X be an arbitrary topological space, $\mathbb{T}$ some measure chain and $t \in \mathbb{T}$ left-dense. If for a mapping $g : \mathbb{T} \to X$ the limit

$$\lim_{s \to t} g(s) = \lim_{s \to t, s < t} g(s) = g(t^-)$$

exists, then we call it the left-sided limit of g at t. Accordingly, we define the notion of the right-sided limit $g(t^+)$ for right-dense points t.

The mapping g is called regulated, if in each left-dense $t \in \mathbb{T}$ the left sided, and in each right-dense $t \in \mathbb{T}$ the right sided limit exists.

Finally, the mapping g is called rd-continuous, if it

(a) is continuous in each right-dense or maximal $t \in \mathbb{T}$, and

(b) the left sided limit $g(t^-)$ exists in each left-dense t.

By $C_{rd}[\mathbb{T}, X]$ we denote the set of rd-continuous mappings $\mathbb{T} \to X$.

The following implications are immediate

$continuous \Rightarrow rd - continuous \Rightarrow regulated.$

If $\mathbb{T}$ contains at the same time left-dense and right-scattered points only, we call them ldrs-points. Then the first implication is not invertible. On a discrete measure chains all three notions coincide.

Theorem 1.4.1: *Let the mapping* $g : \mathbb{T} \to X$ *be given.*

(i) *If* g *is regulated or rd-continuous, resp., then the Bernoulli-shift* $g.\sigma$ *has this property, too.*

(ii) *Let* Y *be a topological space and* $h : X \to Y$ *be continuous. If* g *is regulated or rd-continuous, rep., then so is* $h.g$.

(iii) *If the mappings* $g_n : \mathbb{T} \to X$ *of the sequence* $(g_n)_{n \in N}$ *are regulated or rd-continuous and locally uniformly convergent to* $g : \mathbb{T} \to X$, *then* g *is regulated or rd-continuous, resp., too.*

(iv) *For* $i = 1, \ldots n$ *let* X_i *be some topological space and let the mapping* g_i *be regulated or rd-continuous, resp. Then so is* $g :=(g_1, \ldots, g_n) : \mathbb{T} \to X_1 \times \ldots \times X_n$.

It follows that the graininess $\mu^*(.) = \mu^*(\sigma(.), r) - \mu^*(., r)$ *is rd-continuous. In ldrs-points (e.g. in the points* $k + \frac{1}{2}$, $k \in Z$, *or* P), *however, it is not continuous.*

Proof: For (i) it is enough to consider the one sided limits. Let t be right-dense or left-dense, resp., and let $(s_n)_{n \in N}$ be some sequence with $s_n > t$ or $s_n < t$, resp., converging to t. Then $(r_n)_{n \in N}$, $r_n = \sigma(s_n)$, is a sequence with the same properties. It follows that

$$(g.\sigma)(t \pm) = \lim_{n \to \infty} g.\sigma(s_n) = \lim_{n \to \infty} g(r_n) = g(t \pm).$$

If t is right-dense and g rd-continuous, then we have

$$(g.\sigma)(t\pm) = g(t\pm) = g(t) = (g.\sigma)(t).$$

For (ii), (iii) and (iv) we only mention the fact that limits and the respective operations under considerations commute.

Theorem 1.4.2: *Let* $\tau \in \mathbb{T}$, $x \in X$ *and a regulated mapping* $g : \mathbb{T}^k \to X$ *be given. Then there exists exactly one function, the so-called pre-antiderivative, which is pre-differentiable and fulfills the identities*

$$f^{\Delta}(t) = g(t) \text{ for } t \in D \text{ and } f(\tau) = x.$$

Proof: The uniqueness immediately follows from Corollary 1.3.1 (ii). Let some fixed $n \in N$ be given. For $t \in [\tau, \infty[$ we prove by induction the statements:

$A(t)$ There exists a function $f_{n,t} : [\tau, t] \to X$ which is pre-differentiable with $D_{n,t}$ such that

$$f_{n,t}(\tau) = x \text{ and } \left|f_{n,t}^{\Delta}(s) - g(s))\right| \leq \tfrac{1}{n} \text{ for } s \in D_{n,t}.$$

(I) We set $D_{n,t} = \phi$ and $f_{n,t}(\tau) = x$.

(II) Let t be right-scattered, the function $f_{n,t}$ and its domain of differentiation are assumed to be constructed. We define

$$D_{n,\sigma(t)} = D_{n,t} \cup \{t\} \text{ and}$$

$$f_{n,\sigma(t)} : [\tau, \sigma(t)] \to X$$

$$s \to \begin{cases} f_{n,t}(s)\,,\ if\ s \in [\tau, t] \\ f_{n,t}(t) + g(t)\,\mu^*(t),\ if\ s = \sigma(t). \end{cases}$$

Then we have (cf. 1.2.7),

$$|f^{\Delta}_{n,\sigma(t)}(s) - g(s)\,| \begin{cases} \le \frac{1}{n}\ if\ s \in D_{n,t} \quad , \\ = 0,\ if\ s = t. \end{cases}$$

and hence $A(\sigma(t))$ is true.

(III) Let t be right-dense. By the induction we assume that $D_{n,t}$ and $f_{n,t}$ are given. Then there is a neighborhood U of t such that

$$|g(s) - g(t^+)\,| \le \tfrac{1}{n}\,, \quad s \in U,\ s > t.$$

For $r \in U,\ r > t,$ define

$$D_{n,r} = (D_{n,t} \backslash \{t\}) \cup\,]t, r]^k \ ,$$

$$f_{n,r} : [\tau, r] \to X$$

$$s \to \begin{cases} f_{n,t}(s), \quad if\ s \in [\tau, t] \\ f_{n,t}(t) + g(t^+)\,\mu\,(s, t), \quad if\ s \in\,]t, r]. \end{cases}$$

Then $f_{n,r}$ is pre-differentiable with $D_{n,r}$; by (1.2.6) we get for $s \in\,]t, r\,]^k$

$$|f^{\Delta}_{n,r}(s) - g(s)\,| = |g(t^+) - g(s)\,| \le \tfrac{1}{n}.$$

Together with the induction condition $A(t)$ this yields $A(r)$, $r \in U$, $r > t$.

(IV) Let t be left-dense, for each $r < t$, $D_{n,r}$ and $f_{n,r}$ are assumed to be given. Then there exists some neighborhood U of t such that

$$|g(s) - g(t^-)| \leq \tfrac{1}{n}, \quad s \in U, \ s < t.$$

Fix some $r \in U$, $r < t$. We define:

$$D_{n,r} := \begin{cases} D_{n,r} \cup \,]r,t[, & \text{if r is right-dense} \\ D_{n,r} \cup [r,t[, & \text{if r is right-scattered} \end{cases}$$

and

$$f_{n,t} : [\tau, t] \to X$$

$$s \to \begin{cases} f_{n,r}(s), & if \ s \in [\tau, r] \\ f_{n,r}(t) + g(t^-)\,\mu\,(s,r), & if \ s \in \,]r,t]. \end{cases}$$

Then $f_{n,t}$ is pre-differentiable with $D_{n,t}$ and we have for $s \in D_{n,t}$, $s \geq r$

$$|f_{n,t}^{\Delta}(s) - g(s)| = |g(t^-) - g(s)| \leq \tfrac{1}{n}.$$

and hence $A(t)$ is true.

By means of dual considerations one can prove $A(t)$ for $t \leq \tau$, too. For each fixed $n \in N$, we then have found a function $f_n : \mathbb{T} \to X$ with domain of definition D_n such that

$$|f_n^{\Delta}(t) - g(t))| \leq \tfrac{1}{n} \text{ for } t \in D_n.$$

we have $f_n(\tau) = x$ for all $n \in N$. Theorem 1.3.2 yields a function f which is pre-differentiable with $D = \bigcap_{n \in N} D_n$ such that

$$f^{\Delta}(t) = g(t), \quad t \in D.$$

We do not want to derive an integral notion by means of measure theoretical considerations. we introduce integration as the inverse of differentiation. For a regulated function $g : \mathbb{T}^k \to X$, let $f : \mathbb{T} \to X$ be the pre-antiderivative given in Theorem 1.4.2 . By

$$\int_r^s g(t)\Delta t: = f(s) - f(r) \in X, \tag{1.4.1}$$

we define the cauchy-integral from r to s of the function g.

In the case (R) each regulated real valued function is Reimann-integrable, too. The Riemann-integral coincides with the Cauchy-integral. The advantage of the Cauchy-integral is that it is defined for functions with values in an arbitrary Banach space.

For (hZ) one easily proves the relation

$$\int_r^s g(t)\Delta t = \begin{cases} \sum_{i=\frac{r}{h}}^{\frac{s}{h}-1} g(ih), & \text{if } s > r \\ 0, & \text{if } s = r \\ -\sum_{i=\frac{s}{h}}^{\frac{r}{h}-1} g(ih), & s < r. \end{cases}$$

For (P) and $r, s \in P$ we have

$$\int_r^s g(t)\Delta t = \int_r^{[r]} g(t)\, dt + \sum_{k=[r]}^{[s]-1} [\int_k^{k+\frac{1}{2}} g(t) dt + \tfrac{1}{2} g(k+\tfrac{1}{2})] + \int_{[s]}^s g(t)\, dt.$$

Here the sum has to be furnished with a minus sign, if $[s]-1 < [r]$ (Gauss – bracket).

Now, the different properties of differentiation can be translated into the integral language.

Theorem 1.4.3:

(i) *Suppose the mappings* f *and* $g : \mathbb{T}^k \rightarrow X$ *are regulated. Then we have for* $r, s \in \mathbb{T}$, $\alpha, \beta \in K$:

$$\int_r^s [\alpha f(t) + \beta g(t)]\Delta t = \alpha \int_r^s f(t)\Delta t + \beta \int_r^s g(t)\Delta t. \tag{1.4.2}$$

(ii) *Partial integration: In the situation of Theorem 1.2.3* (ii) *the mappings* $f : \mathbb{T}^k \rightarrow X_1$, *and* $g : \mathbb{T} \rightarrow X_2$ *are assumed to be differentiable with regulated derivatives* f^Δ *and* g^Δ. *Then the stated integrals exist and for* $r, s \in \mathbb{T}$:

$$\int_r^s f(\sigma(t)) g^\Delta(t)\Delta t + \int_r^s f^\Delta(t) g(t)\Delta t = f(s)g(s) - f(r)g(r) \tag{1.4.3}$$

In the case (Z) *this is called Abel summation.*

(iii) *Let the mappings* $f : \mathbb{T}^k \rightarrow X$, $g : \mathbb{T}^k \rightarrow R_0^+$ *be regulated. If*

$$|f(t)| \leq g(t) \text{ for } t \in [r, s]^k$$

then,

$$|\int_r^s f(t)\Delta t| \leq \int_r^s g(t)\Delta t \qquad (1.4.4)$$

(iv) *If the sequence* $(g_n)_{n\in N}$ *of regulated functions* $\mathbb{T}^k \to X$ *converges uniformly on* $[r, s)$ *to the regulated function* g, *then the sequence*

$$(\int_r^s g_n \Delta t)_{n\in N} \text{ converges to } \int_r^s g(t)\Delta t \text{ in } X.$$

Proof: We only have to show that the integrands are regulated. This, however, follows immediately from Theorem 1.4.1.

It is well-known from continuous analysis that the mapping

$$t \to \int_r^t g(s)\, ds \qquad (1.4.5)$$

is differentiable in $t \in R$ with derivative $g(t)$, if g is continuous in t. We already know that (1.4.5) is possibly not differentiable in the right-dense points of $\mathbb{T}^k \backslash D$. Therefore, if one wants to guarantee the differentiability of (1.4.5), it remains to postulate the continuity of g only for these points. We arrive at the notion of rd-continuity as if described before.

Let g be a mapping $\mathbb{T}^k \to X$. The mapping $f : \mathbb{T} \to X$ is called antiderivative of g on $\mathbb{T}$, if it is differentiable on $\mathbb{T}$ and satisfies

$$f^{\Delta}(t) = g(t), \quad \text{for } t \in \mathbb{T}^k.$$

Theorem 1.4.4: *If $g : \mathbb{T}^k \to X$ is rd-continuous, then g has the antiderivative*

$$f : t \to \int_{\tau}^{t} g(s)\,\Delta s.$$

Proof: Suppose $t \in \mathbb{T}$. From the definition of the integral and Theorem 1.4.2 it follows that f is pe-differentiable with D such that $f^{\Delta}(t) = g(t), t \in D$. Hence, let $t \in \mathbb{T}^k \backslash D$, then t is right-dense and $\sigma(t) = t$. For $\epsilon > 0$, there exists a neighborhood U of t such that,

$$|g(s) - g(t)| \leq \epsilon \text{ for all } s \in U.$$

The function $h(.) = f(.) - g(t)\,\mu\,(.,\tau),, \tau \in \mathbb{T}$ fixed, is pre-differentiable with D, we have for $s \in U \cap D$.

$$|h^{\Delta}(s)| = |g(s) - g(t)| \leq \epsilon.$$

With Corollary 1.3.1, we get for $r \in U$

$$|f(t) - f(r) - g(t).\mu.(t,r)|$$

$$= |[f(t) - g(t).\mu.(t,t)] - [f(r) - g(t).\mu.(r,t)]|$$

$$= |h(t) - h(r)|$$

$$\leq \sup_{s \in U^k \cap D} \{|(h_t^{\Delta}(s)|\}|\mu.(t,r)| \leq \epsilon.|\mu.(t,r)|$$

Hence f has a derivative in t. It is uniquely determined by Theorem 1.2.2 (i).

1.5 Notes

The contents of Chapter 1, which form the core of the rest of the chapters, and develops systematically the necessary calculus on time scales, is taken from Hilger [18, 19] and Aulbach and Hilger [5, 6]. Note that the induction argument employed by Dieudonne [11] to prove basic results in differential and integral calculus, is adopted for proving the calculus on time scales suitably.

CHAPTER 2

2.0 Introduction

This chapter introduces dynamic systems on time scales and proves the basic properties of solutions of such systems so that the theory developed includes the corresponding theories of differential equations and difference equations at the same time. The approach brings out clearly the extra assumptions needed when we deal with difference equations directly. As we shall see the induction principle is employed very crucially in most of the proofs.

Section 2.1 considers the existence and uniqueness results for dynamic systems on a time scale under Perron's type of uniqueness condition. Dynamic inequalities are discussed in Section 2.2 for systems, which are then utilized to prove existence of extremal solutions in Section 2.3. Comparison results are proved in Section 2.4, where generalized inner product is also used.

In Section 2.5, we investigate linear dynamic systems on time scales and obtain linear variation of parameters formula. Continuous dependence with respect to initial data as well as a parameter, is treated in Section 2.6. Section 2.7 is devoted to the development of nonlinear variation of parameters formula. For this purpose, we first prove the differentiability results relative to the initial values of solutions, and then employing this result, we obtain the nonlinear variation of parameters formula, which is a complicated expression. Finally, in Section 2.8, we study, as an application of comparison principle, a global existence result and a simple stability criteria.

2.1 Local Existence and Uniqueness

In this section, we shall consider the initial value problem for dynamic systems on time scales and prove local existence and uniqueness results corresponding to Peano's and Perron's theorems [34]. Let $\mathbb{T}^k$ represent the set of all nondegenerate points of the time scale $\mathbb{T}$.

Consider the initial value problem (IVP)

$$x^{\Delta} = f(t, x), t \in \mathbb{T}^k, x(t_0) = x_0, \tag{2.1.1}$$

where $f: \mathbb{T}^k \times R^n \to R^n$ and f is rd-continuous on $\mathbb{T}^k \times R^n$.

A map $x: \mathbb{T}^k \to R^n$ is a solution of (IVP) (2.1.1) if $x(t)$ is an antiderivative of $f(t, x(t))$ on $\mathbb{T}^k$ and satisfies $x(t_0) = x_0$.

Theorem 2.1.1: *Let $f \in C_{rd}[R_0, R^n]$ where $R_0 = [t_0, t_0 + a] \times B$, $[t_0, t_0 + a]$ is understood as $[t_0, t_0 + a] \cap \mathbb{T}^k$ and $B = \{x \in R^n: |x - x_0| \leq b\}$. Then the IVP (2.1.1) has at least one solution $x(t)$ on $[t_0, t_0 + \alpha]$ where $\alpha = min(a, \frac{b}{M})$, M being the bound of $f(t, x)$ on R_0.*

Proof: For any $r \in \mathbb{T}^k$, $t_0 < r < t_0 + \alpha$, define the mapping

$$f^{r]}(t, x) = \begin{cases} f(t, x), & t \in [t_0, r), \quad x \in B, \\ f(r^-, x), & t = r, \quad x \in B. \end{cases}$$

Let the statement $A(r)$ be as follows:

The IVP

$$x^{\Delta} = f^{r]}(t, x),\ t \in [t_0, r],\ x(t_0) = x_0, \tag{2.1.1.r}$$

has a solution $x_r(t)$ on $[t_0, r]$.

(I) The statement $A(t_0)$ is trivially true since the mapping

$$x_{t_0}: \{t_0\} \to B \text{ and } x_{t_0}^{\Delta}(t) = f^{t_0]}(t, x_{t_0}(t)) \text{ for } t \in \{t_0\}^k = \emptyset.$$

(II) Let r be right-scattered and $A(r)$ be true i.e. the IVP (2.1.1.r) has a solution $x_r(t)$ on $[t_0, r]$. Define the mapping $x_{\sigma(r)}$: $[t_0, \sigma(r)] \to B$ such that

$$x_{\sigma(r)}(t) = \begin{cases} x_r(t), & t \in [t_0, r], \\ x_r(r) + f(r, x_r(r))\mu^*(r), & t = \sigma(r). \end{cases}$$

This $x_{\sigma(r)}$ is continuous and is a solution of (2.1.1.r) on $[t_0, \sigma(r)]$.

(III) Let r be right-dense and U_r be a neighborhood of r. Assume $A(r)$ is true. We need to prove that $A(s)$ is true for $s \in U_r \cap [t_0, t_0 + \alpha]$, $s > r$. By classical existence theorem (Peano's theorem) [34] there exists a solution $x_s(t)$ satisfying

$$x_s^{\Delta}(t) = f(t, x_s(t)), t \in [r, s], s \in U_r \cap [t_0, t_0 + \alpha],$$
$$x_s(r) = x_r(r).$$

The mapping defined by

$$x_s(t) = \begin{cases} x_r(t), & t \in [t_0, r] \\ x_s(t), & r \le t \le s, \quad s \in U_r \cap [t_0, t_0 + \alpha] \end{cases}$$

is a solution of $(2.1.1.r)$ on $[t_0, s]$, $s \geq r$, proving $A(s)$ is true.

(IV) Let r be left-dense such that $A(s)$ is true for all $s < r$. We need to prove that $A(r)$ is true. For any $s < r$, the IVP $(2.1.1.r)$ has a solution $x_s(t)$ on $[t_0, s]$ defined by

$$x_s(r) = x_0 + \int_{t_0}^{t} f(\tau, x_s(\tau)) \Delta \tau, t \in [t_0, s]$$

Since $f(t,x)$ is rd-continuous, $\lim_{t \to r^-} f(t, x_s(t))$ exists and hence we have

$$x_s(r) = x_0 + \int_{t_0}^{r} f(\tau, x_s(\tau)) \Delta \tau.$$

Thus $x_s(t)$ is a solution of $(2.1.1.r)$ on $[t_0, r]$ i.e. $A(r)$ holds.

By the induction principle IVP (2.1.1) has a solution on $[t_0, t_0 + \alpha]$ and the proof is complete.

Next we shall consider Perron type uniqueness result.

Theorem 2.1.2: *Assume that*

(i) $g \in C_{rd}[[t_0, t_0 + a] \times [0, 2b], R_+]$ *and for every* t_1, $t_0 \leq t_1 \leq t_0 + a$, $u(t) \equiv 0$ *is the only solution of*

$$u^{\Delta} = g(t, u), u(t_1) = 0,$$

on $[t_1, t_0 + a]$;

(ii) $f \in C_{rd}[\, R_0, R^n \,]$ *and for each* $t \in [\, t_0, t_0 + a \,]$, *there exists a compact neighborhood* U_t *such that* $f^{t]}$ *in* $U_t \times B$ *satisfies*

$$| \, f(t,x) - f(t,y) \, | \;\leq g(t, \; | \, x - y \, | \;), (t,x), (t,y) \in U_t \times B.$$

Then the IVP (2.1.1) *has a unique solution* $x(t)$ *on* $[t_0, t_0 + a]$.

Proof: We apply the induction principle to the following statements

$$A(r)\text{: The IVP } x^{\Delta} = f^{r]}(t,x),\ t \in [\, t_0, r]x(t_0) = x_0, \qquad (2.1.2r)$$

admits exactly one solution $x_r(.)$.

(I) In fact there exists only one mapping $x_{t_0} : \{t_0\} \to R^n$ with $x_{t_0}(t_0) = x_0$ and

$$x^{\Delta}_{t_0}(t) = f^{r]}(t, x_{t_0}(t)) \text{ for } t \in \{t_0\}^k = \phi.$$

(II) Let r be right-scattered. The IVP (2.1.2r) has - according to the induction condition - exactly one solution $x_r(.)$. We define the mapping

$$x_{\sigma(r)} : [t_0, \sigma(r)] \to R^n \text{ by :}$$

$$x_{\sigma(r)}(t) = \begin{cases} x_r(t) \;, \; if\; t \in [t_0, r] \\ x_r(r) + f(r, x_r(r))\mu^*(r), \;\text{ if } t = \sigma(r). \end{cases}$$

It is continuous and the only solution of the IVP (2.1.2$\sigma(r)$), since its restriction to $[t_0, r]$ is the only solution of the IVP (2.1.2r) and its restriction to $[r, \sigma(r)]$ is the only solution of the IVP

$$x^{\Delta} = f(t, x), \quad x(r) = x_r(r) \qquad \text{on } [r, \sigma(r)].$$

(III) Let r be right-dense. By the induction condition there exists exactly one solution $x_r(.)$ of (2.1.2r). Let $V_r \subseteq U_r$ be a compact neighborhood of r. By Perron's Theorem for each $s \in V_r$, $s \geq t$ the IVP

$$x^{\Delta} = f^{s]}(t, x), \ \ t \in [r, s], \ \ x(r) = x_r(r)$$

admits exactly one solution $y_s(.)$. The mapping x_s, defined by

$$x_s(t) := \begin{cases} x_r(t), \ if \ t \in [t_0, r\,] \\ y_s(t), \ \ if \ t \in [r, s\,] \end{cases}, \tag{2.1.2s}$$

is the unique solution of the IVP (2.1.2s). Hence we have $A(s)$ for all $s \in V_r$, $s \geq r$.

(IV) Let r be left-dense, and choose V_r as above, then there is a $s \in V_r$ with $s < r$.

With the help of the induction condition $A(s)$ and Perron's Theorem, existence and uniqueness of a solution $x_r(.)$ of (2.1.2r) can be shown exactly in the same way as in (III). Hence we have $A(r)$.

Since there is a solution on each interval $[t_0, r]$, $r \geq t_0$, there is one on $[t_0, t_0 + a]$, and the proof is complete.

2.2 Dynamic Inequalities

In this section, we shall prove basic results on dynamic inequalities, that are needed to prove existence of extremal solutions. We shall first prove a result relative to a system of strict dynamic inequalities and then consider a similar result for nonstrict inequalities which is needed for later discussion. All inequalities between vectors are to be understood componentwise hereafter.

We need the following definition before we proceed further

Definition 2.2.1: A function $f \in C[R^n, R^n]$ is said to be quasimonotone nondecreasing if

$x \leq y$ and $x_i = y_i$ for some $1 \leq i \leq n$ implies $f_i(x) \leq f_i(y)$.

Theorem 2.2.1: *Let $\mathbb{T}$ be the time scale with $t_0 \geq 0$ minimal element and no maximal element , $v,w:\mathbb{T} \to R^n$ be the rd-continuous mappings that are differentiable for each $t \in \mathbb{T}$ and satisfy*

$$v^{\Delta}(t) \leq f(t,v(t)), \qquad w^{\Delta}(t) > f(t,w(t)) \qquad t \in \mathbb{T}, \tag{2.2.1}$$

where $f \in C_{rd}[\mathbb{T} \times R^n, R^n]$, $f(t,x)$ is quasimonotone nondecreas-ing in x, and, for $1 \leq i \leq n$, $f_i(t,x)\mu^*(t) + x_i$ is nondecreasing in x_i for each $t \in \mathbb{T}$. Then

$$v(t_0) < w(t_0) \quad \text{implies} \quad v(t) < w(t), \text{ for } t \in \mathbb{T}.$$

Proof: We apply the induction principle to the statement

$$A(t) : v(t) < w(t), \quad t \in \mathbb{T}$$

(I) $A(t_0)$ is clearly satisfied since $v(t_0) < w(t_0)$

(II) Let t be right-scattered and $A(t)$ be true. We shall show that $A(\sigma(t))$ is true. Using definition of the derivative for right scattered point, we get

$$v(\sigma(t)) - w(\sigma(t)) = [v^{\Delta}(t) - w^{\Delta}(t)]\mu^*(t) + v(t) - w(t),$$

which, because of (2.2.1) and the fact that $f(t,x)\mu^*(t) + x$ is nondecreasing in x, reduces to

$$v(\sigma(t)) - w(\sigma(t)) = [f(t,v(t)) - f(t,w(t))]\mu^*(t) + v(t) - w(t) < 0$$

since *A(t)* is true. Hence, A(σ(t)) is true.

(III) Let t be right-dense and N be a neighborhood of t. Assume that A(t) is true.

We need to show that *A(s)* is true for $s \geq t$, $s \in N$.

But this follows directly from the theory of differential inequalities for systems see [34].

(IV) Let t be left-dense such that *A(s)* is true for $s < t$. We need to show that *A(t)* is true.

By rd-continuity of v and w, it follows that

$$v(t) = \lim_{s \to t^-} v(s) \leq \lim_{s \to t^-} w(s) = w(t).$$

It remains to show that $v(t) = w(t)$, $t \in \mathbb{T}$ is not possible. Assume, on the contrary that $v(t) = w(t)$. Then,

$$v^{\Delta}(t) - w^{\Delta}(t) < f(t, v(t)) - f(t, w(t)) = 0. \qquad (2.2.2)$$

Using the definition of the derivative , we obtain, for $1 < j < n$,

$$v_j^{\Delta}(\sigma(t)) - v_j(s) - \tfrac{\epsilon}{2} \mid \sigma(t - s \mid \; \leq v_j^{\Delta}(t)\,(\sigma(t) - s)$$

$$- \, w_j^{\Delta}(\sigma(t)) + w_j(s) = \tfrac{\epsilon}{2} \mid \sigma(t) - s \mid \; \leq \; - \, w_j^{\Delta}(t)\,(\sigma(t) - s)$$

which yield

$$(v_j^{\Delta}(t) - w_j^{\Delta}(t))(\sigma(t) - s)$$

$$\geq (v_j^{\Delta}(\sigma(t)) - w_j^{\Delta}(\sigma(t))) - (v_j(s) - w_j(s))$$

Since, $\sigma(t) - s \geq t - s > 0$, $A(s)$ is true, and $v(t) = w(t)$ we arrive at

$$v_j^{\Delta}(t) - w_j^{\Delta}(t) > 0$$

which contradicts (2.2.2). Hence by the induction principle we conclude that $v(t) < w(t)$, $t \in \mathbb{T}$. This completes the proof of the theorem.

The next result deals with nonstrict dynamic inequalities

Theorem 2.2.2: *Let* $\mathbb{T}$ *be the time scale as before,* $v, w: \mathbb{T} \to R^n$ *be the rd- continuous mappings that are differentiable for each* $t \in \mathbb{T}$ *and satisfy*

$$v^{\Delta}(t) \leq f(t, v(t)), \; w^{\Delta}(t) \geq f(t, w(t)), \quad t \in \mathbb{T}.$$

where $f \in C_{rd}[\mathbb{T} \times R^n, R^n]$, $f(t, x)$ *is quasimonotone nonde-creasing in* x *and for each* i, $1 \leq i \leq n$, $f_i(t, x)\mu^*(t) + x_i$ *is nondecreasing in* x_i *for* $t \in \mathbb{T}$. *Then* $v(t_0) \leq w(t_0)$ *implies* $v(t) \leq w(t)$, $t \in \mathbb{T}$, *provided* f *satisfies*

$$f_i(t, x) - f_i(t, y) \leq L \sum_{i=1}^{n} (x_i - y_i), \quad x \geq y. \tag{2.2.3}$$

Proof: We apply the induction principle to the statement

$$A(t)\colon v(t) \le w(t), \quad t \in \mathbb{T}.$$

(I) $A(t_0)$ is clearly satisfied since $v(t_0) \le w(t_0)$.

(II) Let t be right-scattered and $A(t)$ be true. We shall show that $A(\sigma(t))$ is true.

Using the definition we get as before,

$$v(\sigma(t)) - w(\sigma(t)) = [v^{\Delta}(t) - w^{\Delta}(t)]\mu^*(t) + v(t) - w(t),$$

which, because of (2.2.2) and the fact that $f(t,x)\mu^*(t) + x$ is nondecreasing in x, reduces to

$$v(\sigma(t)) - w(\sigma(t)) \le [f(t,v(t)) - f(t,w(t))]\mu^*(t) + v(t) - w(t)$$
$$\le 0,$$

since $A(t)$ is true. Hence $A(\sigma(t))$ is true.

(III) Let t be right dense and N be a neighborhood of t. Assume that $A(t)$ is true. We need to show that $A(s)$ is true for $s \ge t$, $s \in N$. Consider $w_0(t) = w(t) + \ \epsilon e^{(n+1)L(s-t)}$ where ϵ is the n-dimensional vector $(\epsilon, \epsilon, \ldots, \epsilon)$, $\epsilon > 0$. We shall show that $v(s) < w_0(s)$ for $s \ge t$, $s \in N$. If this is not true, there would exist a $s_0 > t$, $s_0 \in N$ and an index j, $1 \le j \le n$ such that

$$v_j(s_0) = w_{0j}(s_0), v_i(s_0) \le w_{0i}(s_0), \ i \ne j$$
$$\text{and } v_i(s) < w_{0i}(s), t \le s \le s_0$$

This implies that $v_j^{\Delta}(s_0) \geq w_{0j}^{\Delta}(s_0)$ and consequently, using quasimonotone property of f and (2.2.3), we arrive at

$$f_j(s_0, v_1(s_0), \ldots, v_j(s_0), \ldots v_n(s_0))$$
$$\geq f_j(s_0, w_1(s_0), \ldots, w_j(s_0), \ldots, w_n(s_0))$$
$$- (n+1)L\epsilon e^{(n+1)L(s-s_0)},$$
$$\geq f_j(s_0, w_{01}(s_0), \ldots, w_{0j}(s_0), \ldots, w_{0n}(s_0)) - L\epsilon e^{(n+1)L(s-s_0)}$$
$$\geq f_j(s_0, v_1(s_0), \ldots, v_j(s_0), \ldots, v_n(s_0) - L\epsilon e^{(n+1)L(s-s_0)}.$$

This contradiction proves $v(s) < w_0(s),\ s \in N$. Since $\epsilon > 0$ is arbitrary, we obtain letting $\epsilon \to 0$, $v(s) \leq w(s),\ s \in N,\ s \geq t$ and therefore $A(s)$ is true.

(IV) Let t be left dense such that $A(s)$ is true for $s < t$. We need to show that $A(t)$ is true. By rd-continuity of v and w, it follows that $v(t) = \lim_{s \to t^-} v(s) \leq \lim_{s \to t^-} w(s) = w(t)$, proving $A(t)$ is true.

Thus by the induction principle, we conclude that $v(t) \leq w(t)$, $t \in T$ and the proof is complete.

2.3 Existence of extremal solutions

Using the result on strict dynamic inequalities proved in section 2.2, we shall discuss, in this section, the existence of extremal solutions for dynamic systems. For that consider the IVP

$$u^{\Delta} = g(t,u) \qquad u(t_0) = u_0 > 0 \tag{2.3.1}$$

where $g \in C_{rd}[R_0, R^n]$, $R_0 = \{([t_0, t_0 + a] \bigcap \mathbb{T}) \times B\}$,

$B = \{u \in R^n: |u - u_0| \leq b\}$, and $|\, g(t,u) \,| \leq M$ on R_0.

Theorem 2.3.1: *Assume that*

(i) *$g(t,u)$ is quasimonotone nondecreasing in u,*

(ii) *for each i, $1 \leq i \leq n$, $g_i(t,u)\mu^*(t) + u_i$ is nondecreasing in u_i for each $t \in \mathbb{T}$.*

Then there exist minimal and maximal solutions of (2.3.1) *on* $I \equiv [t_0, t_0 + \eta] \bigcap \mathbb{T}$ *where* $\eta = \min(a, \frac{b}{2M+b})$.

Proof: Let $0 < \epsilon < b/2$ and consider the following initial value problem

$$u^{\Delta} = g(t,u) + \epsilon, \quad u(t_0,\epsilon) = u_0 + \epsilon \tag{2.3.2}$$

Since $g(t,u) + \epsilon$ is defined and rd-continuous on $R_\epsilon = \{[t_0, t_0 + a]\bigcap \mathbb{T})$ $\times B_\epsilon\}$ where $B_\epsilon = \{u \in R^n : |u - (u_0+\epsilon)| \leq b/2\}$ and $R_\epsilon \subseteq R_0$, we conclude from the local existence theorem on time scale that (2.3.2) has a solution $u(t, \epsilon)$ on the interval I.

For $0 < \epsilon_2 < \epsilon_1 \leq \epsilon$, we have

$$u^\Delta(t,\epsilon_2) \leq g(t,u(t,\epsilon_2))+ \epsilon_2,$$

$$u^\Delta(t,\epsilon_1) > g(t,u(t,\epsilon_1))+ \epsilon_2,$$

$$u(t_0,\epsilon_1) > u(t_0,\epsilon_2).$$

Now by application of Theorem 2.2.1 we obtain $u(t,\epsilon_2) < u(t,\epsilon_1)$ for $t \in$ I.

Since $|g(t,u)| \leq M$ on R_0, the family of functions $\{u(t,\epsilon_n)\}_{n>0}$ is equicontinuous and uniformly bounded on I . Therefore, using standard arguments, it can be shown that $\{u(t,\epsilon_n)\}_{n>0}$ converges uniformly on I to an rd-continuous function, say $r(t)$. Hence,

$$\lim_{n \to \infty} u(t,\epsilon_n) = r(t) \quad \text{with } r(t_0) = u_0.$$

Since $g(t,u)$ is *rd*-continuous on I term by term integration is possible which implies that $r(t) = (r_1(t) \ \ r_2(t) \ldots. \ r_n(t))$ *is* a solution of (2.3.2).

To show that $r(t)$ is the desired maximal solution of (2.3.2) on I, we have to prove that if $u(t)$ is any solution of (2.3.2) on I, then $u(t) \leq r(t)$ on I. For any $t \in$ I and $\epsilon \leq b/2$ we have

$$u(t_0) = u_0 \leq u_0 + \epsilon \equiv u(t_0, \epsilon)$$

$$u^{\Delta}(t) \leq g(t,u)$$

$$u^{\Delta}(t,\epsilon) > g(t,u(t,\epsilon)).$$

Hence, by Theorem 2.2.1 we conclude that

$$u(t) < u(t,\epsilon), \quad t \in \mathrm{I}.$$

Taking the limit on each side as $\epsilon \to 0$ we have

$$u(t) \leq \lim_{\epsilon \to 0} u(t,\epsilon) = r(t), \; t \in \mathrm{I}.$$

The existence of a minimal solution can be proven following similar arguments. Hence, the proof of the theorem is complete.

2.4 Comparison results

Having the existence results and the theory of dynamic inequalities at our disposal, it is now easy to prove comparison results.

Theorem 2.4.1: *Let the assumptions of Theorem 2.3.1 hold and let $m\colon I \equiv [t_0, t_0 + a) \bigcap \mathbb{T} \to R^n$ be a mapping that is differentiable for each $t \in I$ and that satisfies*

$$m^{\Delta}(t) \leq g(t, m(t)), \qquad t \in I.$$

Then $m(t_0) \leq u_0$ implies that

$$m(t) \leq r(t), \qquad t \in I,$$

where $r(t)$ is the maximal solution of (2.3.1) existing on I.

Proof: Let

$$u(t) = g(t,u) + \epsilon, \epsilon > 0, u(t_0) = u_0 + \epsilon \tag{2.4.1}$$

Since $g_\epsilon(t,u) \equiv g(t,u) + \epsilon$ is defined and rd-continuous on $R_\epsilon = \{I \times B_\epsilon\}$ where $B_\epsilon = \{u \in R^n : |u - (u_0 + \epsilon)| \leq b/2\}$ and $R_\epsilon \subseteq R_0$, we conclude by Theorem 2.4.1 that (2.4.1) has a maximal solution $r(t,\epsilon)$ on I where $\eta = min(a, \frac{b}{2M+b})$. For $\epsilon > 0$,we have

$$u^{\Delta}(t,\epsilon) = g(t, u(t,\epsilon)) + \epsilon > g(t, u(t,\epsilon))$$

$$m^{\Delta}(t) \leq g(t, m(t))$$

and

$$u(t_0,\epsilon) = u_0 + \epsilon > u_0$$

$$m(t_0) \leq u_0 < u(t_0,\epsilon).$$

Hence, by Theorem 2.2.1 it follows that

$$m(t) < r(t,\epsilon)\,, \quad t \in I,$$

and $\lim\limits_{\epsilon \to 0} r(t,\epsilon) = r(t)$ uniformly on I. Hence, $m(t) \le r(t)$ for $t \in I$.

Sometimes, it is more convenient to have comparison results in terms of the generalized derivative of $|x|$ and therefore we shall prove such a result next. For this purpose, we need the following definition.

Definition 2.4.1: Let $x \in C^1_{rd}[\mathbb{T}, R^n]$. Given an $\epsilon > 0$, if there exists a neighborhood N_ϵ of $t \in \mathbb{T}$ satisfying

$$\frac{1}{\mu\,(t,s)}\,[|x(\sigma(t))| - |x(\sigma(t)) - \mu\,(t,s)\; x^\Delta(t)\,|] < [x, x^\Delta]_+ + \epsilon$$

for each $s \in N_\epsilon$ and $s > t$, where $\mu\,(t,s) = \sigma(t)$-s, then we say that $[x, x^\Delta]_+$ is the generalized derivative of $x(t)$. In case, $t \in \mathbb{T}$ is rs, then we have

$$[x, x^\Delta]_+ = \frac{1}{\mu^*\,(t)}[\,|x(\sigma(t))| - |x(t)|]$$

where $\mu^*\,(t) = \mu\,(t,t)$.

We can now prove the following comparison result.

Theorem 2.4.2: *Suppose that*

$$[x, x^{\Delta}]_{+} \leq g(t, |x|) \text{ on } \mathbb{T} \times R$$

where $g \in C_{rd}[\mathbb{T} \times R_{+}, R]$ *and* $g(t,u)\,\mu^*(t) + u$ *is nondecreasing in* u *for each* $t \in \mathbb{T}$, *where* $x : \mathbb{T} \to R^n$ *is any rd-continuously differentiable function such that* $|x_0| \leq u_0$. *Then*

$$|x(t)| \leq r(t), \quad t \in \mathbb{T},$$

where $r(t) = r(t, t_0, u_0)$ *is the maximal solution of (2.4.3) existing on* $\mathbb{T}$.

Proof: Define $m(t) = |x(t)|$ and note that $m(t_0) \leq u_0$. We now apply the induction principle to the statement

$$A(t) : \; m(t) \leq r(t), \; t \in \mathbb{T}.$$

(I) $A(t_0)$ is valid since $|x_0| \leq u_0$.

(II) Let $t \in \mathbb{T}$ be rs and $A(t)$ be true. We have to show that $A(\sigma(t))$ is true.

By definition, we have

$$m(\sigma(t)) - r(\sigma(t)) = [D^+ m^{\Delta}(t) - r^{\Delta}(t)]\,\mu^*(t) + m(t) - r(t)$$

$$\leq g(t, m(t)) - g(t, r(t)) \leq 0,$$

in view of the assumptions $g(t,u)\,\mu^*(t) + u$ is nondecreasing and $A(t)$ is true.

But $D^+m^\Delta(t) = \frac{m(\sigma(t))-m(t)}{\mu^*(t)} = \frac{|x(\sigma(t))|-|x(t)|}{\mu^*(t)} = [x(t), x^\Delta(t)]_+$

$\leq |x^\Delta(t)|$, and hence, it follows $A(\sigma(t))$ is true, in either case.

(III) Let t be rd, $A(t)$ be true and let N be the right neighborhood of $t \in \mathbb{T}$. We need to show that $A(s)$ is true for $s \in N$ and $s > t$. In both cases we easily obtain

$$D^+m^\Delta(s) \leq g(s, m(s)), \, t \leq s, \, s \in N,$$

and consequently, by the comparison theorem for differential equations, we get

$$m(s) \leq r(s), \, t \leq s, \, s \in N,$$

proving the claim $A(s)$ is true since $A(t)$ is valid. This conclude the proof, since we are interested only to the right of $t \in \mathbb{T}$ and hence (IV) of induction principle is not needed.

(IV) Let t be left-dense such that $A(s)$ is true for $s < t$. We need to show that $A(t)$ is true. By rd-continuity of the functions involved, we get

$$\lim_{s \to t^-} m(s) \leq \lim_{s \to t^-} r(s),$$

which implies that $A(t)$ is valid. Hence by the induction principle, $A(t)$ holds good for all $t \in \mathbb{T}$ and the proof is complete.

2.5 Linear Variation of Parameters

In contrast to differential equations the solutions of difference equations are in general, not uniquely determined in backward time. So we have to introduce a condition which guarantees unique backward continuation of solutions.

Let ($\mathbb{T}$, μ , X) a dynamical triple, and $B(X)$ be a Banach algebra with unity of the continuous endomorphims on a Banach space X.

A mapping $A : \mathbb{T}^k \to B(X)$ is called regressive, if for each $t \in \mathbb{T}^k$ the mapping $A(t)\mu^*(t) + id : X \to X$ is invertible. This is the case e.g. if $|A(t)\mu^*(t)| < 1$ for all $t \in \mathbb{T}$. Obviously in case $\mathbb{T} = R$ any A is regressive (since μ^*=0) and in case $\mathbb{T} = Z$, A is regressive if $|A(t)| < 1$ (since $\mu^* \equiv 1$).

Suppose $A : \mathbb{T}^k \to B(X)$ is rd-continuous and regressive and $F : \mathbb{T}^k \times X \to X$ is rd-continuous, then a mapping $x : \mathbb{T}^k \to X$ is called a solution of the dynamic equation

$$x^{\Delta} = A(t)\, x + F(t, x) \tag{2.5.1}$$

if $x^{\Delta}(t) = A(t)\, x(t) + F(t, x(t))$ for all $t \in \mathbb{T}^k$.

If a solution $x(.)$ of (2.5.1) in addition satisfies the condition $x(\tau) = \eta$ for a pair $(\tau, \eta) \in \mathbb{T}^k \times X$, it is called a solution of the initial value problem (IVP)

$$x^{\Delta} = A(t)x + F(t,x)\,,\; x(\tau) = \eta\,. \tag{2.5.2}$$

Consider the IVP, in the Banach algebra $B(X)$,

$$x^{\Delta} = A(t)\,x,\;\; x(\tau) = I, \tag{2.5.3}$$

(where I is the unity of $B(X)$). By Theorem 2.1.2 , it admits exactly one solution $\Phi_A(\tau) :\,= x(.;\tau, I)$.

We call it principal solution. The corresponding transition function is defined to be $\Phi_A(t,\tau) :\,= \Phi_A(\tau)(t)$. In the particular case, when $A : \mathbb{T}^k \to B(X)$ is constant , we call the transition function, exponential function ($e_L(t,\tau)$) .

If $X = R$ and $C : \mathbb{T}^k \to R_+$, then $C(t)\mu^*(t) + 1 > 0$ satisfies the regressive property we can set $e_C(t,\tau) :\,= \Phi_A(t,\tau)$ and then it is clear because of Theorems 2.4.1 and 2.5.1, the following result which is the Gronwall's inequality in the present set up.

Lemma 2.5.1: *Let a, b, c be continuous functions from $\mathbb{T}^k \to R_+$. Then the inequality*

$a(t) \leq b(t) + \int_\tau^t c(s)\, a(s) \Delta s$ for all $t \in \mathbb{T}^k$,

implies that

$a(t) \leq b(t) + \int_\tau^t c(s) e_c(t, \sigma(s))\, b(s) \Delta s$ for all $t \in \mathbb{T}^k$.

Lemma 2.5.2: *If* a *is a real constant with* $a\,\mu^*(t) + 1 > 0$ *for all* $t \in \mathbb{T}^k$, *then one gets* $e_a(t, \tau) \geq 1 + \mu(t, \tau)$ *for all* $t, \tau \in \mathbb{T}^k$.

Remarks: 1. For (R) this is the obvious estimate

$$e^{a(t-\tau)} \geq 1 + a(t-\tau),$$

involving a Taylor series truncation and for (hZ), $h > 0$ one gets Bernoulli's inequality

$$(ha + 1)^{(t-\tau)/h} \geq 1 + a(t - \tau).$$

2. The innocent looking estimate from lemma 2.5.2 is of crucial importance for the unified qualitative treatment of continuous and discrete time dynamics.

Theorem 2.5.1: *We consider the IVP* (2.5.2) *with rd-continuous and regressive right-hand side. Then the solution of* (2.5.2) *is given by*

$x(t) = \Phi_A(t, \tau)\, \eta + \int_\tau^t \Phi_A(t, \sigma(s))\, F(s, x(s)) \Delta s$

Proof: By defining $x(t) = \Phi_A(t,\tau)\, z(t)$ and substituting in (2.5.2), we obtain using product rule (Theorem 1.2.3(*ii*)),

$$\Phi_A(\sigma(t),\tau)\, z^\Delta + \Phi_A^\Delta(t,\tau)\, z = A(t)\Phi_A(t,\tau)\, z + F(t,x),$$

which yields,

$$\begin{aligned} z^\Delta &= \Phi_A^{-1}(\sigma(t),\tau)\, F(t,x(t)) \\ &= \Phi_A(\tau,\sigma(t))\, F(t,x(t)). \end{aligned}$$

Thus $z(t) = \eta + \int_\tau^t \Phi_A(t,\sigma(s))\, F(s,x(s))\ \Delta\, s.$

Multiplying this equation by $\Phi_A(t,\tau)$ gives

$$\begin{aligned} x(t) &= \Phi_A(t,\tau)\,\eta + \int_\tau^t \Phi_A(t,\tau)\Phi_A(\tau,\sigma(s))\, F(s,x(s))\ \Delta\, s \\ &= \Phi_A(t,\tau)\,\eta + \int_\tau^t \Phi_A(t,\sigma(s))\, F(s,x(s))\ \Delta\, s. \end{aligned}$$

2.6 Continuous dependence

In this section, we shall discuss continuous dependence of solutions of dynamic systems on time scales with respect to initial values as well as parameters. We need such results later for proving differentiability of solutions relative to initial values.

Theorem 2.6.1: *Let* $\mathbb{T}=[\tau, s]$ *be some compact measurable chain and* L *be a negative constant with* $L\ \mu(s,\tau) < 1$. *The right-hand side in the IVP*

$$x^{\Delta} = f(t,x), \ \ x(\tau) = \eta, \tag{2.6.1}$$

is rd-continuous and satisfies the Lipschitz condition

$$|f(t,x_1) - f(t,x_2)\ | \leq L\,|x_1 - x_2|, \ t \in \mathbb{T}^k, \, x_1, x_2 \in R^n$$

Then the solutions $x(.,\tau,\eta)$ *of* (2.6.1) *are unique and continuous with respect to the initial values* (τ,η).

Proof: Since the uniqueness of solutions follow from Theorem 2.1.2, we have to prove the continuity part only.

Let $x(t,\tau,\eta)$, $x(t,\tau_1,\eta_1)$ be the solutions through (τ,η) , (τ_1,η_1) respectively. Define

$v(t)=|x(t,\tau,\eta) \ - \ x(t,\tau_1,\eta_1)\ |$, then $v(t) = \ |\eta \ - \ x(\tau,\tau_1,\eta_1)\ |$.

We claim that

$$v(t) \leq m(t), \text{ for all } t \geq \tau$$

where $m(t) = e_L(t,\tau)\, v\,(\tau)$ is a solution of IVP in R,

$$u^{\Delta} = L\,u, \ \ u(\tau) = v(\tau).$$

We apply the induction principle Theorem 1.1.1 to the following

$$A(t) : v(t) \leq m(t), \text{ for all } t \geq \tau.$$

(I) Obviously, $A(\tau)$ is true.

(II) Let t be right-scattered and $A(t)$ be true.

We need to show that $A(\sigma(t))$ is true. By Theorem 2.1.2 (v), we get

$$v(\sigma(t)) - m(\sigma(t))$$
$$= [v^{\Delta}(t) - m^{\Delta}(t)]\, \mu^*(t) + v(t) - m(t) \tag{2.6.2}$$

Since

$$v_t^{\Delta}(t) = \frac{v(\sigma(t)) - v(t)}{\mu^*(t)}$$

$$= \frac{|x(\sigma(t),\tau,\eta) - x(\sigma(t),\tau_1,\eta_1)| - |x(t,\tau,\eta) - x(t,\tau_1,\eta_1)|}{\mu^*(t)}$$

$$\leq |x^{\Delta}(t,\tau,\eta) - x^{\Delta}(t,\tau_1,\eta_1)|$$

$$= |f(t, x(t,\tau,\eta)) - f(t, x(t,\tau_1,\eta_1))|$$

$$\leq L v(t).$$

Therefore, in view of (2.6.2), we obtain

$$v(\sigma(t)) - m(\sigma(t)) = L\,[v(t) - m(t)]\, \mu^*(t) + v(t) - m(t)$$
$$\leq 0,$$

since $A(t)$ is valid.

Hence $A(\sigma(t))$ is true.

(III) Let t be right-dense and U be a neighborhood of t. Suppose that $A(t)$ is true. As in the theory of ordinary differential equations [34], one can get

$$v(s) \le m(s), \; s > t, \; s \in U.$$

Hence $A(s)$ is satisfied for all $s \in U, s > t$.

(IV) Let t be left-dense such that $A(s)$ is true for all s with $s < t$. By continuity of v and m, it follows that

$$v(t) = \lim_{s \to t^-} v(s) \le \lim_{s \to t^-} m(s) = m(t).$$

Hence $A(t)$ is valid.

Thus $A(t)$ is true for all $t \ge \tau$.

Let $t \in \mathbb{T}$ be fixed and U be a compact neighborhood of τ containing t.

We define

$$M := max\{\, e_L(t,s) : \; s \in U\} > 0.$$

For $\epsilon > 0$, there exists a neighborhood of τ such that for all $s \in V$,

$$\left| x(s,\tau,\eta) - x(\tau,\tau,\eta) \right| \le \tfrac{\epsilon}{2M}.$$

For $s \in U \cap V$ and $y \in N_{\frac{\epsilon}{2M}}$ (a neighborhood of η),

$$\begin{aligned} \left| x(t,s,y) - x(t,\tau,\eta) \right| &\le e_L(t,s) \left| y - x(s,\tau,\eta) \right| \\ &\le e_L(t,s) \left[\left| y - \eta \right| + \left| x(s,\tau,\eta) - x(\tau,\tau,\eta) \right| \right] \end{aligned}$$

$$\leq M\left[\frac{\epsilon}{2M} + \frac{\epsilon}{2M}\right] = \epsilon$$

The proof is complete.

Theorem 2.6.2: *Let* $\mathbb{T}= [\tau, s]$ *be some compact measure chain and* L *be a nonnegative constant with* $L\ \mu(s,\tau) < 1$. *Let* $f \in C_{rd}[\mathbb{T}^k \times D \times A, R^n]$, D *is an open set in* R^n, *and* A *is an open parameter* μ_0 *– set in* R^m, *and for* $\mu = \mu_0$, *let* $x_0(t) = x(t, \tau, \eta, \mu_0)$ *be a solution of*

$$x^\Delta = f(t, x, \mu_0),\ \ x(\tau) = \eta \tag{2.6.3}$$

existing for $t \geq \tau$. *Assume further that*

$$|f(t, x, \mu) - f(t, x, \mu_0)| < \eta,$$

provided $\mu \in U_{\mu_0}$ *(a neighborhood of* μ_0*), and*

$$|f(t, x, \mu) - f(t, x, \mu_0)| \leq L\,|x_1 - x_2|,\ \ t \in \mathbb{T}^k$$

Then, given $\epsilon > 0$, *there exists a neighborhood* U_{μ_0} *of* μ_0 *such that for every* $\mu_1 \in U_{\mu_0}$, *the IVP*

$$x^\Delta = f(t, x, \mu_1),\ \ x(\tau) = \eta \tag{2.6.4}$$

admits a unique solution $x(t) = x(t, \tau, \eta, \mu_1)$ *satisfying*

$$|x(t) - x_0(t)| < \epsilon,\ t \geq \tau.$$

Proof: The uniqueness of solution is obvious from Theorem 2.1.2 .

Consider the IVP ,

$$u^{\Delta} = Lu + \eta, \quad u(\tau) = 0 \qquad (\text{here } \eta = \eta(\epsilon), \ \epsilon > 0).$$

By using variation of parameter formula

$$u(t, \tau, 0, \eta) = \int_{\tau}^{t} e_L(t, \sigma(s))\, \eta \,\Delta s$$

$$\leq \eta \sup_{s \in [\tau, t]^k} e_L(t, \sigma(s))\,(t - \tau).$$

Let $\epsilon > 0$ and let V be a compact neighborhood of t. Then

$$u(t, \tau, 0, \eta) < \epsilon, \; t \in V.$$

Now, define

$$m(t) = |x(t) - x_0(t)|,$$

where $x(t)$, $x_0(t)$ are the solutions of (2.6.3), (2.6.4) respectively. Then as in the proof of Theorem 2.6.1, we can prove that

$$m(t) \leq u(t, \tau, 0, \eta), \; t \in \mathbb{T} \text{ whenever } \mu_1 \in U_{\mu_0}.$$

Hence

$$|x(t) - x_0(t)| < \epsilon, \; t \in V, \text{ provided that } \mu_1 \in U_{\mu_0}.$$

which completes the proof of the theorem.

2.7 Nonlinear Variation of Parameters

To obtain nonlinear variation of parameters formula, analogous to Alekseev's formula for dynamic systems on time scales, we require the differentiability of solutions relative to initial data. we shall therefore begin with such a result. For this purpose, we need the following result.

Lemma 2.7.1: *Let $f \in C_{rd}[\mathbb{T}^k \times D, R^n]$, where D is an open , convex set in R^n, and let f_x exist and be rd-continuous. Then,*

$$f(t,x_2) - f(t,x_1) = \int_0^1 f_x(t, sx_2 + (1-s)\,x_1)\,\Delta\, s\ (x_2 - x_1).$$

Proof: Define

$$F(t,s) = f(t, sy + (1-s)x),\ \ 0 \le s \le 1.$$

The convexity of D implies that $F(t,s)$ is well defined. It is obvious that $F(t,s)$ is rd-continuous. Since $f_x(t,x)$ exists, it follows that $F_s^{\Delta}(t,s)$ exists. Moreover its derivative $F_s^{\Delta}(t,s)$ is rd-continuous. Hence

$$F_s^{\Delta}(t,s) = f_x(t, sx_2 + (1-s)x_1).(x_2 - x_1) \tag{2.7.1}$$

since $F(t,1) = f(t,x_2)$ and $F(t,0) = f(t,x_1)$, the result follows by integrating (3.4) , in the sense of Cauchy with respect to s from 0 to 1, which completes the proof.

Theorem 2.7.1: *Let* $\mathbb{T}$= [τ, s] *be some compact measure chain. Assume that* $f \in C_{rd}[\mathbb{T}^k \times R^n, R^n]$, *and possesses rd-continuous partial derivatives* f_x *on* $\mathbb{T}^k \times R^n$.

Let L *be a nonnegative constant with* $L\ \mu(s,\tau) < 1$ *and* $|f_x(t,x)| \le L$ *on* $\mathbb{T}^k \times R^n$. *Let the solution* $x_0(t) = x(t,\tau,\eta)$ *of*

$$x^{\Delta} = f(t,x),\ \ x(\tau) = \eta,$$

exists for $t \ge \tau$.

Then

(i) $\Phi(t,\tau,\eta) = x_{\eta}(t,\tau,\eta)$ *exists and is the solution of*

$$y^{\Delta} = H(t,\tau,\eta)\ y \tag{2.7.2}$$

where

$$H(t,\tau,\eta) = \lim_{h \to 0} \int_0^1 f_x(t, p x(t,\tau,\eta) - (1-p) x(t,\tau,\eta + h))\, \Delta p$$

such that $\Phi(\tau,\tau,\eta)$ *is the unit matrix;*

(ii) $\Psi(t,\tau,\eta) = x_{\tau}^{\Delta}(t,\tau,\eta)$ *exists, is the solution of*

$$z^{\Delta} = H(t,\sigma(\tau),\tau,\eta)\ z \tag{2.7.3}$$

such that $\Psi(\sigma(\tau),\tau,\eta) = -f(\tau,\eta)$, *where*

$$H(t,\sigma(\tau),\tau,\eta)$$

$$= \int_0^1 f_x(t, p x(t,\sigma(\tau),\eta) - (1-p) x(t,\tau,\eta))\, \Delta p;$$

(iii) *the function* $\Phi(t,\tau,\eta), \Psi(t,\tau,\eta)$ *satisfy the relation*

$$\Psi(t,\tau,\eta) = -\Phi(t,\sigma(\tau),\eta)\, f(\tau,\eta)$$

$$+\int_{\sigma(\tau)}^{t} \Phi(t,\sigma(s),\eta)\,[H(s,\sigma(\tau),\tau,\eta)$$

$$-H(s,\sigma(\tau),\tau,\eta)]\Psi(s,\tau,\eta)\,\Delta s$$

Proof: Under the assumptions on f, it is clear , by Theorem 2.1.2 and Theorem 2.6.1, that the solutions $x(t,\tau,\eta)$ exist, are unique and continuous in (t,τ,η) on some interval.

First we shall prove (ii). Let $\epsilon > 0$ and let U be a neighborhood of τ .By Lemma 2.7.1, we have for $s \in U$,

$$[x(t,\sigma(\tau),\eta) - x(t,s,\eta)]_t^{\Delta}$$

$$= f(t,x(t,\sigma(\tau),\eta)) - f(t,x(t,s,\eta))$$

$$= \int_0^1 f_x(t,px(t,\sigma(\tau),\eta)$$

$$+(1-p)x(t,s,\eta))\,\Delta p\,[x(t,\sigma(\tau),\eta) - x(t,s,\eta)].$$

Thus we get IVP of the type

$$y_t^{\Delta}(t) = H(t,\sigma(\tau),s,\eta)\,y(t), \tag{2.7.4}$$

$$y(\sigma(\tau)) = -\frac{1}{\mu(\sigma(\tau),s)}\int_s^{\sigma(\tau)} f(\xi,x(\xi,\tau,\eta))\,\Delta\xi,$$

(here $\mu(\sigma(\tau),s) = \sigma(\tau) - s$)

where

$H(t,\sigma(\tau),s,\eta)=\int_0^1 f_x(t,px(t,\sigma(\tau),\eta)+(1-p)x(t,s,\eta))\ \Delta p.$

If τ is right-dense, that is , $\sigma(\tau)=\tau$, then we get desired result as in the theory of ordinary differential equation.

If τ is right-scattered, then consider $\frac{x(t,\sigma(\tau),\eta)-x(t,s,\eta)}{\mu(\sigma(\tau),s)}$, we know that

$$\left|\frac{x(t,\sigma(\tau),\eta)-x(t,s,\eta)}{\mu(\sigma(\tau),s)}-\frac{x(t,\sigma(\tau),\eta)-x(t,\tau,\eta)}{\mu^*(t)}\right|\leq\epsilon\ ,$$

since the mapping $g(.):=\frac{x(t,\sigma(\tau),\eta)-x(t,.,\eta)}{\mu(\sigma(\tau),.)}$ is defined and continuous in a neighborhood of τ.

Now,

$$\left|x(t,\sigma(\tau),\eta)-x(t,s,\eta)-\frac{x(t,\sigma(\tau),\eta)-x(t,\tau,\eta)}{\mu^*(t)}\mu(\sigma(\tau),s)\right|$$

$$\leq\epsilon\left|\mu(\sigma(\tau),s)\right|.$$

It further implies that

$$x_\tau^\Delta=\frac{\partial x(t,\tau,\eta)}{\partial\tau}=\frac{x(t,\sigma(\tau),\eta)-x(t,\tau,\eta)}{\mu^*(\tau)}\text{ exists.}$$

We now show that x_τ^Δ is the solution of (2.7.3).

By using Lemma 2.7.1, we get

$$\frac{\partial x(\sigma(\tau),\tau,\eta)}{\partial\tau}=\frac{x(\sigma(\tau),\sigma(\tau),\eta)-x(\sigma(\tau),\tau,\eta)}{\mu^*(\tau)}$$

$$= \frac{\eta - x(\sigma(\tau),\tau,\eta)}{\mu^*(\tau)}$$

$$= -\frac{1}{\mu^*(\tau)} \int_\tau^{\sigma(\tau)} x_\tau^\Delta(\xi,\tau,\eta)\,\Delta\xi$$

$$= -\frac{1}{\mu^*(\tau)} \int_\tau^{\sigma(\tau)} f(\xi, x(\xi,\tau,\eta))\,\Delta\xi$$

$$= -f(\tau,\eta).$$

The proof of (i) follows immediately from the theory of ordinary differential equation, using the rd-continuous nature of f_x.

Finally, to prove (iii), we proceed as follows.

Since $\Phi(t,\ \sigma(\tau),\eta) = \frac{\partial x(t,\sigma(\tau),\eta)}{\partial\eta}$ exists and is the solution of

$$y^\Delta = H(t,\sigma(\tau),\eta)\,y,$$

where

$$H(t,\sigma(\tau),\eta) = \lim_{h\to 0} \int_0^1 f_x(t, p\,x(t,\sigma(\tau),\eta)$$

$$-(1-p)x(t,\sigma(\tau),\eta+h))\,\Delta p$$

such that

$\Phi(\sigma(\tau),\ \sigma(\tau),\eta)$ is the unit matrix.

But

$$\Psi_t^\Delta = H(t,\sigma(\tau),\eta)\Psi + [H(t,\sigma(\tau),\tau,\eta) - H(t,\sigma(\tau),\eta)]\,\Psi$$

Hence, by variation of parameter formula (Theorem 3.1), we have

$$\Psi(t,\tau,\eta) = -\Phi(t,\ \sigma(\tau),\eta)\,f(\tau,\eta)$$

$$+ \int_{\sigma(\tau)}^{t} \Phi(t, \sigma(s), \eta)[H(s, \sigma(\tau), \tau, \eta)$$

$$- H(s, \sigma(\tau), \eta)] \Psi(s, \tau, \eta) \Delta s.$$

which completes the proof.

Finally, we are now in a position to prove the nonlinear variation of parameter formula.

Theorem 2.7.2: *Let* $\mathbb{T} = [\tau, s]$ *be some compact measure chain. Assume that* $f, F \in C_{rd}[\mathbb{T}^k \times R^n, R^n]$, *and let* f_x *exists and be rd-continuous on* $\mathbb{T}^k \times R^n$.

Let L *be a nonnegative constant with* $L\ \mu(s, \tau) < 1$ *and* $|f_x(t, x)| \leq L$ *on* $\mathbb{T}^k \times R^n$. *If* $x(t, \tau, \eta)$ *is the solution*

$$x^{\Delta} = f(t, x), \quad x(\tau) = \eta,$$

exists for $t \geq \tau$, *any solution* $y(t, \tau, \eta)$ *of* $y^{\Delta} = f(t, y) + F(t, y)$, *with* $y(\tau) = \eta$, *satisfies the integral equation*

$$y(t, \tau, \eta) = x(t, \tau, \eta) + \int_{\tau}^{t} \Phi(t, \sigma(s), y(s)) F(s, y(s)) \Delta s$$

$$+ \int_{\tau}^{t} \int_{\sigma(s)}^{t} \Phi(t, \sigma(p), y(s))[H(p, \sigma(s), s, y(s))$$

$$- H(p, \sigma(s), y(s))] \Psi(p, s, y(s)) \Delta p \Delta s.$$

Proof. Set $y(t) = y(t, \tau, \eta)$. Then, by using Theorem 1.2.3 and Theorem 2.7.1, we get

$$[x(t,s,y(s))]_s^{\Delta} = x_s^{\Delta}(t,s,y(s)) + x_y(t,\sigma(s),y(s))\, y_s^{\Delta}$$

$$= -\Phi(t,\ \sigma(s),\ y(s))f(s,y(s))$$

$$+ \int_{\sigma(s)}^{t} \Phi(t,\sigma(p),\ y(s))[H(p,\sigma(s),s,y(s))$$

$$- H(p,\sigma(s),y(s))]\,\Psi(p,s,y(s))\,\Delta p$$

$$+ \Phi(t,\ \sigma(s),\ y(s))\, y_s^{\Delta}.$$

$$= -\Phi(t,\ \sigma(s),\ y(s))[y_s^{\Delta} - f(s,y(s))]$$

$$+ \int_{\sigma(s)}^{t} \Phi(t,\sigma(p),\ y(s))[H(p,\sigma(s),s,y(s))$$

$$- H(p,\sigma(s),y(s))]\,\Psi(p,s,y(s))\,\Delta p.$$

$$= \Phi(t,\sigma(s),\ y(s))F(s,y(s))$$

$$+ \int_{\sigma(s)}^{t} \Phi(t,\sigma(p),y(s))[H(p,\sigma(s),s,y(s))$$

$$- H(p,\sigma(s),y(s))]\,\Psi(p,s,y(s))\,\Delta p.$$

Since $x(t,t,y(t,\tau,\eta)) = y(t,\tau,\eta)$, by integrating above equation we obtain

$$y(t,\tau,\eta) = x(t,\tau,\eta) + \int_{\tau}^{t} \Phi(t,\ \sigma(s),\ y(s))\, F(s,y(s))\,\Delta s$$

$$+ \int_{\tau}^{t} \int_{\sigma(s)}^{t} \Phi(t,\sigma(p),\ y(s))[H(p,\sigma(s),s,y(s))$$

$$- H(p,\sigma(s),y(s))]\,\Psi(p,s,y(s))\,\Delta p\,\Delta s.$$

Remarks: It is easy to see from the definitions of $H(p, \sigma(s), s, y(s))$ and $H(p, \sigma(s), , y(s))$ that they are identical if the measure chain is R, and consequently, the foregoing Variation of Parameter formula reduces to the usual Alekseev's formula. Moreover, when the dynamic systems is linear, the last term in the expression of the foregoing formula disappears, and one gets the corresponding linear Variation of Parameters formula for both continuous and discrete systems respectively. In view of the present results, it is clear that the results announced in [26] are not correct.

2.8 Global Existence and Stability

As an application of comparison Theorem 2.4.2, we shall prove, in this section, a global existence result and a simple stability result.

Theorem 2.8.1: *Assume that*

(i) $f \in C_{rd}[\mathbb{T} \times R^n, R^n], g \in C_{rd}[\mathbb{T} \times R_+, R_+], g(t,u)$ *is nondecreasing in* u *for each* $t \in \mathbb{T}$, *where* $\mathbb{T}$ *is the time scale with* $t_0 \geq 0$ *as the minimal element and has no maximal element and*

$$| f(t,x) | \leq g(t, | x |) \textit{ for } (t, | x |) \in \mathbb{T} \times R^n$$

(ii) *the maximal solution $r(t)$ of the scalar IVP*

$$u^{\Delta} = g(t,u), u(t_0) = u_0 \geq 0 \tag{2.8.1}$$

exists on $\mathbb{T}$.

Then the largest interval of existence of any solution $x(t)$ of

$$x^{\Delta} = f(t,x), x(t_0) = x_0, \tag{2.8.2}$$

with $|\, x_0 \,| \leq u_0$ *is* $\mathbb{T}$.

Proof: Let $x(t)$ be any solution of (2.8.2) such that $|\, x_0 \,| \leq u_0$ which exists on $[t_0, \beta)$ for $t_0 < \beta < \infty$, $\beta \in \mathbb{T}$ and the value of β cannot be increased. It is easy to see that a solution can have finite escape time only before left-dense points $t \in \mathbb{T}$, since their neighborhoods contain infinitely many points to the left of t. Consequently, let us suppose that β is a left-dense point of $\mathbb{T}$.

Setting $m(t) = \,|\, x(t) \,|$, we get

$$m_+^{\Delta}(t) \leq \,|\, f(t,x(t)) \,|\, \leq g(t,m(t)), [t_0,\beta),$$

$$m(t_0) \leq u_0,$$

(where $m_+^{\Delta}(t)$ is the right-derivative of $m(t)$).

Then by comparison Theorem 2.4.2, we arrive at

$$| x(t) | \leq r(t), [t_0, \beta), \tag{2.8.3}$$

where $r(t)$ is the maximal solution of (2.8.1). For any $t_1, t_2 \in \mathbb{T}$ such that $t_0 < t_1 < t_2 < \beta$, we have, using non-decreasing nature of $g(t, u)$ and (2.8.3)

$$\begin{aligned} | x(t_2) - x(t_1) | &\leq \int_{t_1}^{t_2} g(s, | x(s) |) ds \\ &\leq \int_{t_1}^{t_2} g(s, r(s)) ds = r(t_2) - r(t_1). \end{aligned} \tag{2.8.4}$$

Since $\lim_{t \to \beta^-} r(t)$ exists and is finite, taking limit as $t_1, t_2 \to \beta^-$ and using Cauchy criterion for convergence, it follows from (2.8.4) that $\lim_{t \to \beta^-} x(t)$ exists and is finite. Now define $x(\beta) = \lim_{t \to \beta^-} x(t)$ and consider the IVP

$$x^\Delta = f(t, x), x(\beta) = \lim_{t \to \beta^-} x(t).$$

By local existence Theorem 2.2.1, one gets that $x(t)$ can be continued beyond β, contradicting our assumptions. Hence every solution $x(t)$ of (2.8.2) exists on $\mathbb{T}$ and the proof is complete.

To prove a simple stability result, we need the following definition of stability.

Definition 2.8.1: The trivial solution of (2.7.2) is said to be

(i) stable if given an $\epsilon > 0$ and $t_0 \in \mathbb{T}$, there exists a $\delta > 0$ such that $|x_0| \leq \delta$ implies $|x(t)| \leq \epsilon$, $t \geq t_0$;

(ii) asymptotically stable if it is stable and $\lim_{t \to \infty} |x(t)| = 0$.

We are now in a position to prove a typical result on stability in terms of comparison principle.

Theorem 2.8.2: *Assume that*

(i) $f \in C_{rd}[\mathbb{T} \times R^n, R^n]$, $g \in C_{rd}[\mathbb{T} \times R_+, R]$, $f(t,0) \equiv 0$, $g(t,0) \equiv 0$ *and for* $(t,x) \in \mathbb{T} \times R^n$,

$$[x, f(t,x)]_+ \equiv \lim_{h \to 0^+} \frac{1}{h} \left[|x + hf(t,x)| - |x| \right] \leq g(t, |x|); \tag{2.8.5}$$

(ii) $g(t,u)\mu^*(t) + u$ *is nondecreasing in* u *for each* $t \in \mathbb{T}$.

Then the stability properties of the trivial solution of the IVP

$$u^\Delta = g(t,u), \quad u(t_0) = u_0 \geq 0 \tag{2.8.6}$$

imply the corresponding stability properties of the trivial solution of

$$x^\Delta = f(t,x), \quad x(t_0) = x_0, \quad t_0 \geq 0, \, t \in \mathbb{T}. \tag{2.8.7}$$

Proof: Let the trivial solution of (2.8.6) be stable. Then, given $\epsilon > 0$ and $t_0 \geq 0$, there exists a $\delta = \delta(t_0, \epsilon) > 0$ such that

$$u \leq u_0 < \delta \text{ implies } u(t) < \epsilon, t \in \mathbb{T}. \tag{2.8.8}$$

It is easy to claim that with these ϵ and δ, the trivial solution of (2.8.7) is also stable. If this were false, there would exist a solution $x(t)$ of (2.8.7) with $| x_0 | < \delta$ and a $t_1 \in \mathbb{T}$, $t_1 > t_0$ such that

$$\epsilon \leq | x(t_1) | \text{ and } | x(t) | \leq \epsilon, t \in [t_0, t_1).$$

For $t \in [t_0, t_1]$, using condition (2.8.5), we get

$$m^{\Delta}(t) \leq g(t, m(t)), t \in [t_0, t_1], m(t_0) \leq u_0,$$

where $m(t) = | x(t) |$. Consequently, Theorem 2.4.2 yields

$$| x(t) | \leq r(t), t \in [t_0, t_1].$$

At $t = t_1$, we arrive at the contradiction

$$\epsilon \leq | x(t_1) | \leq r(t_1) < \epsilon,$$

proving the claim. One can prove similarly other concepts of stability, and we omit the details.

2.9 Notes

Section 2.1 includes the results of Kaymakcalan [22]. see also Hilger [19]. Theorem 2.2.1 is taken from Lakshmikantham and Kaymakcalan [33]. For a scalar version see Kaymakcalan [22]. Theorem 2.3.1 and 2.4.1 is work of Kaymakcalan [22]. Theorem 2.4.2 is new. The results of section 2.5 concerning the linear variation of parameters and Gronwall inequality see Aulbach and Hilger [5]. See also Lakshmikantham, Shahzad and Sivasundaram [43]. The contents of sections 2.6 and 2.7 leading to nonlinear variation of parameters formula and related results are taken from Lakshmikantham, Shahzad and Sivasundaram [43]. Section 2.7 covers the work of Kaymakcalan [22]. In several results the original work quoted has been refined suitably to obtain better and correct results. For a survey of results on dynamic systems on time scales see Kaymakcalan and Leela [25]. See Lakshmikantham and Leela [34] for several corresponding results on continuous dynamic systems. Also see Lakshmikantham and Trigiante [40], Agarwal [1], Kocic and Ladas [28] for the corresponding theory of difference equations.

Chapter 3

3.0 Introduction

As is well known, we can develop qualitative behavior of solution of differential systems as well as difference equations by employing Lyapunov-like functions and the theory of corresponding inequalities.

In this chapter, we shall investigate stability criteria by utilizing Lyapunov's method suitably in the framework of the comparison principle. In Section 3.1, we shall prove necessary comparison results in terms of Lyapunov-like functions using the calculus of functions on time scales appropriately. Section 3.2 is devoted to obtain stability criteria which include representative theorems analogous to Lyapunov's theorems as well as general results exploiting comparison principle. Employing the general comparison result proved in Section 3.1 that connects the solutions of two dynamic systems, we give a typical result in Section 3.3, which can be utilized in the study of perturbed systems to get fruitful conclusions.

The notion of conditionally invariant sets and their stability properties are discussed in Section 3.4, which are then applied to consider the controllability of uncertain systems. Section 3.5 offers stability results in terms of two different measures which include and unify several concepts of stability in one set up. Finally, in Section 3.6, we deal with practical stability considerations where we employ the method of Vector Lyapunov functions that offers a flexible mechanism under weaker assumptions.

3.1 Comparison theorems

Consider the dynamic system

$$x^{\Delta} = f(t, x), \; x(t_0) = x_0, \tag{3.1.1}$$

where $f \in C_{rd}[\mathbb{T} \times R^n, R^n]$, and x^{Δ} denotes the derivative of x with respect to $t \in \mathbb{T}$. We shall assume, for convenience, that the solutions $x(t) = x(t, t_0, x_0)$ of (3.1.1) exist and are unique for $t \geq t_0$.

Definition 3.1.1: Let $V \in C_{rd}[\mathbb{T} \times R^n, R_+]$. Then we define the generalized derivative of $V(t, x)$ relative to (3.1.1) as follows: given $\epsilon > 0$, there exists a neighborhood $N(\epsilon)$ of $t \in \mathbb{T}$ such that

$$\frac{1}{\mu(t,s)}[V(\sigma(t), x(\sigma(t))) - V(s, x(\sigma(t)) - \mu(t,s) f(t, x(t))]$$
$$< D^+V^\Delta(t, x(t)) + \epsilon$$

for each $s \in N(\epsilon)$ and $s > t$, where $\mu(t,s) = \sigma(t) - s$ and $x(t)$ is any solution of (3.1.1).

In case, $t \in \mathbb{T}$ is right scattered and $V(t, x(t))$ is continuous at t, we have

$$D^+V^\Delta(t, x(t)) = \frac{1}{\mu^*(t)}[V(\sigma(t), x(\sigma(t))) - V(t, x(t))]$$

where $\mu^*(t) = \mu(t, t)$.

We are now in a position to prove the following comparison theorem in terms of Lyapunov function $V(t, x)$.

Theorem 3.1.1: *Let $V \in C_{rd}[\mathbb{T} \times R^n, R_+]$, $V(t, x)$ be locally Lipschitzian in x for each $t \in \mathbb{T}$ which is rd, and let*

$$D^+V^\Delta(t, x(t)) \le g(t, V(t, x))$$

where $g \in C_{rd}[\mathbb{T} \times R_+, R_+]$, $g(t,u)\mu^*(t) + u$ *is nondecreasing in* u *for each* $t \in \mathbb{T}$ *and* $r(t) = r(t, t_0, u_0)$ *is the maximal solution of* $u^\Delta = g(t,u)$, $u(t_0) = u_0 \geq 0$, *existing on* $\mathbb{T}$. *Then,* $V(t_0, x_0) \leq u_0$ *implies that* $V(t, x(t)) \leq r(t, t_0, u_0), t \in \mathbb{T}, t \geq t_0$.

Proof: We apply the induction principle of Theorem to the statement

$$A(t) : V(t, x(t)) \leq r(t),\ t \in \mathbb{T},\ t \geq t_0.$$

(I) $A(t_0)$ since $V(t_0, x_0) \leq u_0$.

(II) Let t be rs and $A(t)$ be true. We need to show that $A(\sigma(t))$ is true. By definition, if we set $m(t) = V(t, x(t))$, we see that

$$m(\sigma(t)) - r(\sigma(t)) = [\,D^+ m^\Delta(t) - r^\Delta(t)]\mu^*(t) + m(t) - r(t),$$

which, because of $g(t,u)\ \mu^*(t)$ being nondecreasing in u and $A(t)$ being true, reduces to

$$m(\sigma(t)) - r(\sigma(t)) \leq [g(t, m(t)) - g(t, r(t))]\mu^*(t) + m(t) - r(t) \leq 0.$$

In view of the fact that

$$\frac{m(\sigma(t)) - m(t)}{\mu^*(t)} = \frac{V(\sigma(t), x(\sigma(t))) - V(t, x(t))}{\mu^*(t)},$$

we see that $A(\sigma(t))$ is true.

(III) Let t be rd and N be a right neighborhood of $t \in \mathbb{T}$. We need to show that $A(s)$ is true of $s > t$, $s \in N$. This follows from the comparison theorem for differential equations relative to Lyapunov function. See [34].

(IV) Let t be left-dense and let $A(s)$ be true for all $s < t$. We have to prove that $A(t)$ is true. This follows by rd-continuity of $V(t,x)$, $x(t)$ and $r(t)$ and therefore by induction principle, $A(t)$ is valid for all $t \in \mathbb{T}$. The proof is complete.

Remark 3.1.1: If the inequalities between vectors is understood as componentwise, then Theorem 3.1.1 is valid for $V \in C_{rd}[\mathbb{T} \times R^n, R^N{}_+]$, $g \in C_{rd}[\mathbb{T} \times R_+^N, R_+^N]$, provided $g(t,u)$ is quasimonotone nondecreasing in u and $g(t,u)\mu^*(t) + u$ is nondecreasing in u for each $t \in \mathbb{T}$. The proof requires slight modification since we now need to use comparison result for differential systems relative to Vector Lyapunov functions. See [34] . We therefore omit proving such a result.

We shall next discuss another comparison result which connects the solutions to dynamic systems which can be employed in perturbation theory.

Consider another dynamic system

$$x^\Delta = F(t,x), x(t_0) = x_0 \tag{3.1.2}$$

where $F \in C_{rd}[\mathbb{T} \times R^n, R^n]$. Relative to the system (3.1.1), let us assume that the following assumption (H) holds:

(H) the solutions $y(t, t_0, x_0)$ of (4.3.1) exist for all $t \geq t_0$, unique and rd-continuous with respect to the initial data and $\mid y(t, t_0, x_0) \mid$ is locally Lipschitzian in x_0.

For any $V \in C_{rd}[\mathbb{T} \times R^n, R_+]$ and any fixed $t \in \mathbb{T}$, we define $D^+V^\Delta(s, y(t, s, x(s))$ as follows: given $\epsilon > 0$, there exists a neighborhood $N(\epsilon)$ of $s \in \mathbb{T}$, $t_0 \leq s \leq t$ such that

$$\frac{1}{\mu(s,r)}[V(\sigma(s), y(t, \sigma(s), x(\sigma(s))))$$

$$- V(s, y(t, s, x(\sigma(s))) - \mu(s, r)F(s, x(s))]$$

$$< D^+V(s, y(t, s, x(s))) + \epsilon$$

for each $r \in N(\epsilon)$ and $r > s$. As before, if $s \in \mathbb{T}$ is right-scattered and $V(s, y(t, s, x(s)))$ is continuous at s, then

$$D^+V(s, y(t, s, x(s))) = \frac{1}{\mu^*(s)}[V(\sigma(s), y(t, \sigma(s), x(\sigma(s)))$$

$$- V(s, y(t, s, x(\sigma(s))],$$

with $\mu^*(s) = \mu(s, s)$.

We then have the following general comparison result which includes Theorem 3.1.1 as a special case.

Theorem 3.1.2: *Assume that the assumption* (H) *holds. Suppose that*

(i) $V \in C_{rd}[\mathbb{T} \times R^n, R_+]$, $V(s,x)$ *is locally Lipschitzian in* x *for each* $t \in \mathbb{T}$ *which is rd and for* $t_0 < s \leq t$, $x \in R^n$

$$D^+V^{\Delta}(s, y(t,s,x)) \leq g(s, V(s, y(t,s,x)));$$

(ii) $g \in C_{rd}[\mathbb{T}^k \times R_+, R]$, $g(t,u)\mu^*(t) + u$ *is nondecreasing in* u *for each* $t \in \mathbb{T}$, *and the maximal solution* $r(t, t_0, u_0)$ *of*

$$u^{\Delta} = g(t,u), \quad u(t_0) = u_0 \geq 0, \textit{exists for } t \in \mathbb{T}$$

Then, if $x(t) = x(t, t_0, x_0)$ *is any solution of* (4.3.2) *we have*

$$V(t, x(t, t_0, x_0)) \leq r(t, t_0, u_0), \quad t \in \mathbb{T}, \tag{3.1.3}$$

provided $V(t_0, y(t, t_0, x_0)) \leq u_0$.

Proof: Let $x(t) = x(t, t_0, x_0)$ be any solution of (3.1.2). Set

$$m(s) = V(s, y(t, s, x(s))), t_0 \leq s \leq t,$$

so that

$$m(t_0) = V(t_0, y(t, t_0, x_0)).$$

We now apply induction principle to the statement

$$A(s): V(s, y(t, s, x(s))) \leq r(s, t_0, u_0), \quad t_0 \leq s \leq t.$$

(I) $A(t_0)$ is true since $V(t_0, y(t, t_0, x_0)) \leq u_0$.

(II) Let s be rs and $A(s)$ is true. To show that $A(\sigma(t))$ is true.

By definition, we have, following the proof of Theorem 3.1.1,

$$m(\sigma(s)) - r(\sigma(s)) = [\, D^{+}m^{\Delta}(s) - r^{\Delta}(s)]\mu^{*}(s) + m(s) - r(s) \leq 0,$$

where $r(s) = r(s, t_0, u_0)$. Since

$$\frac{m(\sigma(s)) - m(s)}{\mu^{*}(s)} = \frac{V(\sigma(t), y(t, \sigma(s), x(\sigma(s)))) - V(s, y(t, s, x(s)))}{\mu^{*}(t)},$$

and $A(s)$ is true, it is clear that $A(\sigma(s))$ is also true.

(III) Let s be rd and N be a right neighborhood of $s \in \mathbb{T}$. We have to show that $A(\xi)$ is true for $\xi > s$, $\xi \in N$. As before, this follows from the theory of differential equations. See [34]. Hence $A(s)$ is true for $t_0 \leq s \leq t$ and therefore,

$$V(s, y(t, s, x(s))) \leq r(s, t_0, u_0), \quad t_0 \leq s \leq t.$$

since $m(t) = V(t, x(t))$, the desired estimate (3.1.3) follows by setting $s = t$ and the proof is complete.

Taking $u_0 = V(t_0, y(t, t_0, x_0))$, the inequality (3.1.3) becomes

$$V(t, x(t, t_0, x_0)) \leq r(t, t_0, V(t_0, y(t, t_0, x_0))), t \in \mathbb{T},$$

which shows the connection between the solutions of the systems (3.1.1) and (3.1.2) in terms of the maximal solution of the comparison system.

A number of remarks can now be made:

(1) The trivial function $f(t,y) \equiv 0$ is admissible in Theorem 3.1.2 to yield the estimate (3.1.3) provided $V(t_0,x_0) \leq u_0$. In this case, $y(t,t_0,x_0) = x_0$ and hypothesis (H) is trivially verified. Since $y(t,s,x) = x$, the definition $D^+V^\Delta(s,y(t,s,x))$ reduces to $D^+V(t,x)$, which is the usual definition of generalized derivative of the Lyapunov function relative to the system (3.1.2). Consequently, Theorem 4.3.1 reduces, in this special case, to Theorem 3.1.1.

(2) Suppose that $f(t,y) = A(t)y$ where $A(t)$ is an $n \times n$ *rd*-continuous matrix. The solutions $y(t,t_0,x_0)$ of (3.1.1) then satisfy

$$y(t,t_0,x_0) = \Phi(t,t_0)x_0,$$

where $\Phi(t,t_0)$ is the fundamental matrix solution of $y^\Delta = A(t)y$, with $\Phi(t_0,t_0) = I$ (identity matrix). The assumption (H) is clearly verified. Suppose also that $g(t,u) \equiv 0$. Then (3.1.3) yields

$$V(t,x(t,t_0,x_0)) \leq V(t_0,\Phi(t,t_0)x_0), \quad t \geq t_0, \tag{3.1.4}$$

by taking $V(t_0,y(t,t_0,x_0)) = V(t_0,\Phi(t,t_0)x_0) = u_0 = r(t,t_0,u_0)$. If $V(t,x) = \mid x \mid$, (3.1.4) leads to

$$\mid x(t,t_0,x_0) \mid \; \leq \; \mid \Phi(t,t_0)x_0 \mid, t \geq t_0.$$

(3) Suppose that $f(t,y)$ is nonlinear, $f_y^\Delta(t,y)$ exists and is *rd*-continuous for $(t,y) \in \mathbb{T} \times R^n$. Then, the solutions $y(t,t_0,x_0)$ are differentiable with respect to (t_0,x_0) and by Theorem , we have,

$$\begin{cases} y_{t_0}^{\Delta}(t, t_0, x_0) = & -\Phi(t, t_0, x_0) f(t_0, x_0), \ t \geq t_0, \\ & y_{x_0}^{\Delta}(t, t_0, x_0) = \Phi(t, t_0, x_0), \end{cases} \tag{3.1.5}$$

where $\Phi(t, t_0, x_0)$ is the matrix solution of the variational equation

$$z^{\Delta} = f^{\Delta}(t, y(t, t_0, x_0))z,$$

such that $\Phi(t_0, t_0, x_0) = I$.

If $V(s, x)$ is also assumed to be differentiable, then by (3.1.5), we have, for a fixed t,

$$D^{+}V^{\Delta}(s, y(t, s, x)) \equiv V_s^{\Delta}(s, y(t, s, x)) + V_x^{\Delta}(\sigma(s), y(t, s, x)) \cdot \Phi(t, s, x) \cdot [F(s, x) - f(s, x)]. \tag{3.1.6}$$

The relation (3.1.6) gives an intuitive feeling of $D^{+}V^{\Delta}(s, y(t, s, x))$. If, in addition, $V(t, x) = |x|^2$ and $F(t, x) = f(t, x) + R(t, x)$, (3.1.6) yields

$$D^{+}V^{\Delta}(s, y(t, s, x)) \equiv 2y(t, s, x) \cdot \Phi(t, s, x) \cdot R(s, x)$$

which exhibits how the perturbation term is involved in the computation.

3.2 Stability Criteria

In this section, we shall consider some simple stability results. We list a few definitions concerning the stability of the trivial solution of (3.1.1) which we assume to exit.

Definition 3.2.1: The trivial solution $x = 0$ of (3.1.1) is said to be

$(S1)$ *equi-stable* if, for each $\epsilon > 0$ and $t_0 \in \mathbb{T}$, there exists a positive function $\delta = \delta(t_0, \epsilon)$ that is rd-continuous in t_0 for each ϵ such that $| x_0 | < \delta$ implies $| x(t, t_0, x_0) | < \epsilon$ for $t \geq t_0$;

$(S2)$ *uniformly stable* , if the δ in $(S1)$ is independent of t_0;

$(S3)$ *quasi-equi asymptotically stable*, if for each $\epsilon > 0$ and $t_0 \in \mathbb{T}$, there exist positive $\delta_0 = \delta_0(t_0)$ and $T = T(t_0, \epsilon)$ such that $| x_0 | < \delta_0$ implies $| x(t, t_0, x_0) | < \epsilon$ for $t \geq t_0 + T$;

$(S4)$ *quasi-uniformly asymptotically stable*, if δ_0 and T in $(S3)$ are independent of t_0;

$(S5)$ *equi-asymptotically stable*, if $(S1)$ and $(S3)$ hold simultaneously;

$(S6)$ *uniformly asymptotically stable*, if $(S2)$ and $(S4)$ hold simultaneously.

Corresponding to the definitions $(S1)$ to $(S6)$, we can define the stability notions of the trivial solution $u = 0$ of (3.2.1) below. For example, the trivial solution $u = 0$ of (3.2.1) is eqi-stable if, for each $\epsilon > 0$ and $t_0 \in \mathbb{T}$, there exists a function $\delta_0 = \delta_0(t_0, \epsilon)$ that is rd-continuous in t_0 for each ϵ, such that $u_0 < \delta$ implies $u(t, t_0, u_0) < \epsilon, t \geq t_0$.

We are now in a position to prove a general result which provides sufficient conditions for stability criteria.

Theorem 3.2.1: *Assume that*

(i) $V \in C_{rd}[\mathbb{T} \times R^n, R_+]$, $V(t,x)$ *is locally Lipschitzian in* x *for each right dense* $t \in \mathbb{T}$*;*

(ii) $b(\,|\,x\,|\,) \leq V(t,x) \leq a(\,|\,x\,|\,)$ *for* $(t,x) \in \mathbb{T} \times R^n$ *where* $a, b \in \mathcal{K} = [\sigma \in C[R_+, R_+]: \sigma(0) = 0$ *and* $\sigma(u)$ *is increasing in* $u]$*;*

(iii) $f(t,0) = 0$, $g \in C_{rd}[\mathbb{T} \times R_+, R]$, $g(t,u)\mu^*(t) + u$ *is nondecreasing in* u *for each* $t \in \mathbb{T}$*, and*

$$D^+V^\Delta(t,x) \leq g(t, V(t,x)),\ (t,x) \in \mathbb{T} \times R^n.$$

Then the stability properties of the trivial solution of

$$u^\Delta = g(t,u),\ u(t_0) = u_0 \geq 0 \tag{3.2.1}$$

imply the corresponding stability properties of the trivial solution of (3.1.1).

Proof: Let $\epsilon > 0$ and $t_0 \in \mathbb{T}$ be given. Suppose that the trivial solution of (3.2.1) is equi-stable. Then given $b(\epsilon) > 0$ and $t_0 \in \mathbb{T}$, there exists a $\delta_1 = \delta_1(t_0, \epsilon) > 0$ such that

$$u_0 < \delta_1 \Rightarrow u(t) < b(\epsilon),\ t \in \mathbb{T} \tag{3.2.2}$$

where $u(t) = u(t, t_0, u_0)$ is any solution of (3.2.1). Choose $\delta = \delta(t_0, \epsilon) > 0$ such that

$$a(\delta) < \delta_1. \tag{3.2.3}$$

We claim that if $|x_0| < \delta$, then $|x(t)| < \epsilon$, $t \in \mathbb{T}$, where $x(t) = x(t, t_0, x_0)$ is any solution of (3.2.1). If this is not true, there would exist a $t_1 \in \mathbb{T}$, $t_1 > t_0$ and a solution $x(t) = x(t, t_0, x_0)$ of (3.2.1) satisfying

$$|x(t)| < \epsilon, t_0 \le t < t_1 \text{ and } |x(t_1)| \ge \epsilon. \tag{3.2.4}$$

Setting $m(t) = V(t, x(t))$ for $t_0 \le t \le t_1$ and using condition (iii), we get by Theorem 3.1.1, the estimate

$$V(t, x(t)) \le r(t, t_0, u_0), t_0 \le t \le t_1 \tag{3.2.5}$$

where $r(t, t_0, u_0)$ is the maximal solution of (3.2.1) with $V(t_0, x_0) \le u_0$. Now the relations (3.2.2), (3.2.4), (3.2.5) and the assumption (ii) yield

$$b(\epsilon) \le b(\,|x(t_1)|\,) \le V(t, x(t_1)) \le r(t_1, t_0, u_0) < b(\epsilon),$$

since $u_0 = V(t_0, x_0) \le (\,|x_0|\,) < a(\delta) < \delta_1$ by (3.2.3). This contradiction proves the claim.

Other stability properties may be proved in a similar manner and hence the proof is complete.

Remark 3.2.1: Usually when stability properties for differential systems are proved, one imposes conditions on $V(t,x)$ only in $R_+ \times S(\rho)$, where $S(\rho) = [x \in R^n\colon \mid x \mid \ < \rho]$, because stability notions are of local nature relative to the trivial solution. On the other hand, when we deal with difference equations, we need, either to assume $S(\rho)$ is invariant or work in the entire R^n, since we have no control of how large the solutions grow being discontinuous. To avoid assuming the invariance of $S(\rho)$ for some $\rho > 0$, we have imposed conditions everywhere in Theorem 3.2.1. In the present set up, all we need to concentrate is on the estimate (3.2.8). If t_1 is a non-scattered point, that is, say left dense, we then get $\mid x(t_1) \mid \ = \epsilon$ and if t_1 is left scattered point, $\mid x(t_1) \mid \ \geq \epsilon$ and it may happen that $\mid x(t_1) \mid \ > \rho$ if we work in $S(\rho)$.

We can obtain from Theorem 3.2.1, a result analogous to Lyapunov's first theorem for continuous and discrete cases immediately. This we state as a corollary.

Corollary 3.2.1: *The function* $g(t,u) \equiv 0$ *is admissible in Theorem* 3.2.1 *to yield uniform stability of the trivial solution of* (3.2.1).

A result analogous to Lyapunov's Second theorem cannot be obtained directly from theorem 3.2.1, since when we choose $g(t,u) = -c(u)$, $c \in \mathcal{K}, g(t,u)\mu^*(t) + u$ does not satisfy the monotone condition. However, a minor change in the argument proves such a result.

Corollary 3.2.2: *The choice of the function* $g(t,u) = -c(u)$, $c \in \mathcal{K}$, *in Theorem* 3.2.1 *implies uniform asymptotic stability of the trivial solution of* (3.2.1).

Proof: Since $(S2)$ follows from corollary 3.2.1, we need to prove only $(S4)$. To do it, we set $\epsilon = \rho$, for some $\rho > 0$ and designate by $\delta_0 = \delta(\rho)$ so that we have

$$|x_0| < \delta_0 \text{ implies } |x(t)| < \rho, t \geq t_0,\ t \in \mathbb{T}.$$

If $x(t)$ is any solution of (3.2.1) with $|x_0| < \delta_0$, then by setting

$$L(t,x(t)) = V(t,x(t)) + \int_{t_0}^{t} C[V(s,x(s))]\Delta s,$$

we see that $D^+L^\Delta(t,x(t)) \leq 0$, which yields by Theorem 3.1.1

$$L(t,x(t)) \leq L(t_0,x_0) = V(t_0,x_0),\ t \geq t_0. \tag{3.2.6}$$

Choose $T = T(\epsilon) > 0$, for any given $\epsilon > 0$, to satisfy $T > \frac{a(\delta_0)}{C[b(\delta)]}$, where $\delta = \delta(\epsilon)$ is related to ϵ for $(S2)$. Because of (S2), it is enough to show that there exists a $t^* \in [t_0, t_0 + T]$ such that $|x(t^*)| < \delta$. If this is not true, we should have $|x(t)| \geq \delta$, $t \in [t_0, t_0 + T]$ and hence $V(t, x(t)) \geq b(\delta)$, $t \in [t_0, t_0 + T]$. It then follows from (3.2.6)

$$C[b(\delta)] \leq V(t_0 + T, x(t_0 + T)) + \int_{t_0}^{t_0+T} C[V(s, x(s))]\Delta s,$$
$$\leq V(t_0, x_0) \leq a(|x_0|) \leq a(\delta_0),$$

which contradicts the definition of T in (3.2.7). It therefore follows that $(S4)$ is true and the proof is complete.

Usually, Lyapunov's Second Theorem has the assumption $D^+V^\Delta(t, x(t)) \leq -C_0(|x|)$, $C_0 \in \mathcal{K}$. Since $V(t, x)$ has the upper estimate in (ii), which means $V(t, x)$ is decrescent, it is easy to show that $D^+V^\Delta(t, x(t)) \leq -C(|x|)$, where $C(u) = C_0^{-1}.a(u)$. Of course, one can prove directly with the assumption $D^+V^\Delta(t, x(t)) \leq -C_0(|x|)$ following the proof of Corollary 3.2.2.

When $V(t, x)$ is not assumed to be decrescent, that $V(t, 0) \equiv 0$, we have Marachkov's result for differential equations. We can extend such a result in the present set up.

Theorem 3.2.2: *Assume that* $V \in C_{rd}[\mathbb{T} \times R^n, R_+]$, $V(t,x)$ *is locally Lipschitzian in* x *for each* $t \in \mathbb{T}$ *which is rd,* $b(\mid x \mid) \leq V(t,x)$ *and* $V(t,0) \equiv 0$. *Suppose further that* $D^+V^\Delta(t,x(t)) \leq -C(\mid x \mid)$, *and* $\mid f(t,x) \mid \leq M$ *for* $(t,x) \in \mathbb{T} \times R^n$. *Then the trivial solution is equiasymptotically stable.*

Proof: Let $\epsilon > 0$, and $t_0 \in \mathbb{T}$ be given. Since $V(t,x)$ is rd continuous and $V(t,0) \equiv 0$, it is possible to find a $\delta = \delta(t_0, \epsilon) > 0$ satisfying $V(t_0, x_0) < b(\epsilon)$ iff $\mid x_0 \mid < \delta$.

We claim that $\mid x_0 \mid < \delta$ implies $\mid x(t) \mid < \epsilon$, $t \geq t_0$, the proof of which follows from Theorem 3.1.1 . Now set $\epsilon = \rho$, for some $\rho > 0$ and designate by $\delta_0 = \delta(t_0, \rho)$ so that we have $\mid x_0 \mid < \delta_0$ implies $\mid x(t) \mid < \rho, t \geq t_0$. To prove the Theorem, we need to show that $\liminf_{t \to \infty} \mid x(t) \mid \neq 0$, then there exists a $T > 0$ such that for a given $\eta > 0$, we have $\mid x(t) \mid > \eta$, $t \geq T$. As a result, we get arguing as in Corollary 3.2.2, the estimate $V(t, x(t)) \leq V(t_0, x_0) - \int_T^t C(\mid x(s) \mid)\Delta s, t \geq T$, which yields a contradiction , for large t,

$$0 \leq V(t_0, x_0) - C(\eta)(t - T).$$

Hence $\liminf_{t \to \infty} |x(t)| = 0$. Suppose then $\limsup_{t \to \infty} |x(t)| \neq 0$. Then, given an $\epsilon > 0$, there exist a divergent sequence $\{t_k\}$ such that $|x(t_k)| > \epsilon$. Each $t_k \in \mathbb{T}$ may belong to one of the following cases:

(i) t_k is rs and ls;

(ii) t_k is rs and ld;

(iii) t_k is rd and ls;

(iv) t_k is rd and ld;

without loss of generality, we can assume that there is a divergent subsequence $\{t_i\}$ of $\{t_k\}$ such that all t_i belong to one of the above four cases. In case (i), we have

$$V(\sigma(t_i), x(\sigma(t_i))) \leq V(t_i, x(t_i)) - \mu^*(t_i)C(|x(t_i)|),$$

which yields, by successive application,

$$0 \leq V(\sigma(t_i), x(\sigma(t_i))) \leq V(t_0, x_0) - \sum_{j=1}^{i} \mu^*(t_j)C(|x(t_j)|),$$
$$\leq V(t_0, x_0) - C(\epsilon)\,\eta\, i$$

This leads to a contradiction as $i \to \infty$ since $\mu^*(t_i)$ is constant for each i, say η.

In cases (ii) to (iv), we can find another divergent sequence $\{t_i^*\}$ such that $t_i < t_i^*$ or $t_i^* < t_{i+1}$ satisfying

$$|x(t_i)| = \epsilon, \ |x(t_i^*)| = \tfrac{1}{2}\epsilon, \ \tfrac{1}{2}\epsilon < |x(t)| < \epsilon, \ t \in (t_i, t_i^*) \text{ or}$$

$$|x(t_i^*)| = \tfrac{1}{2}\epsilon, \ |x(t_{i+1})| = \epsilon, \ \tfrac{1}{2}\epsilon < |x(t)| < \epsilon, \ t \in (t_i^*, t_{i+1}) .$$

Since $|f(t,x)| \leq M$, it is easy to find $t_i - t_i^* > \frac{\epsilon}{2M}$ and $t_{i+1} - t_i^* > \frac{\epsilon}{2M}$.

It therefore follows that

$$0 \leq V(t_i^*, x(t_i^*)) \leq V(t_0, x_0) - iC(\tfrac{1}{2}\epsilon)\tfrac{\epsilon}{2M} < 0, \text{ for large } i.$$

This is a contradiction. Similarly, we get contradiction for other case. Hence we have $\lim_{t \to \infty} |x(t)| = 0$ and the proof is complete.

3.3 A Technique in Stability Theory

As an application of Theorem 3.1.2, we shall consider a typical result on stability and asymptotic behavior of solutions of (3.1.2).

Theorem 3.3.1: *Assume that* (H) *holds and* (i) *of Theorem* 3.1.2 *is verified. Suppose that* $g \in C_{rd}[\mathbb{T} \times R_+, R]$, $g(t,u)\mu^*(t) + u$ *is nondecreas-*

ing in u *for each* $t \in \mathbb{T}$, $g(t,0) \equiv 0$ $f(t,0) \equiv 0$, $F(t,0) \equiv 0$ *and for* $(t,x) \in \mathbb{T} \times R^n$,

$$b(\mid x \mid) \leq V(t,x) \leq a(\mid x \mid), \; a, b \in \mathcal{K} \tag{3.3.1}$$

Furthermore, suppose that the trivial solution of (3.1.2) *is uniformly stable and* $u = 0$ *of*

$$u^\Delta = g(t,u), \; u(t_0) = u_0 \geq 0 \tag{3.3.2}$$

is uniformly asymptotically stable. Then, the trivial solution of (3.1.2) *is uniformly asymptotically stable.*

Proof: Let $\epsilon > 0$ and $t_0 \in \mathbb{T}$ be given. The uniform stability of $u = 0$ of (3.3.2) implies that given $b(\epsilon) > 0$, $t_0 \in \mathbb{T}$, there exists a $\delta_1 = \delta_1(\epsilon) > 0$ such that if $u_0 \leq \delta_1$, then

$$u(t, t_0, u_0) < b(\epsilon), t \geq t_0. \tag{3.3.3}$$

Let $\delta_2 = a^{-1}(\delta_1)$. Since $y = 0$ of (3.1.2) is uniformly stable, given $\delta_2 > 0$, $t_0 \in \mathbb{T}$, there exists a $\delta = \delta(\epsilon)$ such that

$$\mid y(t, t_0, x_0) \mid \; < \delta_2, t \geq t_0 \text{ if } \mid x_0 \mid \; < \delta. \tag{3.3.4}$$

We claim that $|x_0| < \delta$ also implies that $|x(t, t_0, x_0)| < \epsilon$, $t \in \mathbb{T}$, where $x(t, t_0, x_0)$ is any solution of (3.1.2). If this is not true, there would exist a solution $x(t, t_0, x_0)$ of (3.1.2) with $|x_0| < \delta$ and a $t_1 > t_0$ such that $|x(t_1, t_0, x_0)| \geq \epsilon$, $t_0 \leq t \leq t_1$. Then, by Theorem 3.1.2, we have

$$V(t, x(t, t_0, x_0)) \leq r(t, t_0, V(t_0, y(t, t_0, x_0))),\ t_0 \leq t \leq t_1.$$

Consequently, by (3.3.1), (3.3.3) and (3.3.4), we get

$$\begin{aligned} b(\epsilon) \leq V(t_1, x(t_1, t_0, x_0)) &\leq r(t_1, t_0, a(\,|\,y(t_1, t_0, x_0)\,|\,)) \\ &\leq r(t_1, t_0, a(\delta_2)) \\ &\leq r(t_1, t_0, \delta_1) < b(\epsilon). \end{aligned}$$

This contradiction proves that $x = 0$ of (3.1.2) is uniformly stable.

To show uniform asymptotic stability, we set $\delta = \delta_0$. Then, from the foregoing argument, we have

$$b(\,|\,x(t, t_0, x_0)\,|\,) \leq V(t, x(t, t_0, x_0)) \leq r(t, t_0, V(t_0, y(t, t_0, x_0)))$$

for all $t \in \mathbb{T}$, if $|x_0| \leq \delta_0$. From this it follows that

$$b(\,|\,x(t, t_0, x_0)\,|\,) \leq r(t, t_0, \delta_1),\ t \in \mathbb{T}$$

which implies the uniform asymptotic stability of $x = 0$ because of the assumption that $u = 0$ of (3.3.2) is uniformly asymptotically stable. Hence the proof is complete.

Setting $F(t,x) = f(t,x) + R(t,x)$ in Theorem 3.3.1, we see that although the unperturbed system (3.1.2) is only uniformly stable, the perturbed system (3.1.2) is uniformly asymptotically stable, an improvement caused by the perturbing term.

3.4 Stability of Conditionally Invariant Sets

Consider the differential system

$$x^{\Delta} = f(t,x,\lambda), \ x(t_0) = x_0 \ t_0 \in \mathbb{T} \tag{3.4.1}$$

where $f \in C_{rd}[\mathbb{T} \times R^m \times R^d, R^n]$ and $\lambda \in R^d$ is a parameter. Also consider the differential equation

$$u^{\Delta} = g(t, u \mid \lambda \mid), \ \ u(t_0) = u_0 \geq 0, t_0 \in \mathbb{T} \tag{3.4.2}$$

where $g \in C_{rd}]\mathbb{T} \times R_+, R]$.

Let us begin by defining the notions of self-invariant and conditionally invariant sets, as well as relevant stability concepts. For that purpose, let $0 < r < \rho$ and

$$A = \{x \in R^n \mid x \mid \ \leq r\} \text{ and } B = \{x \in R^n \colon \mid x \mid \ \leq \rho\}. \tag{3.4.3}$$

Also let $x(t) = x(t, \lambda, t_0, x_0)$ be any solution of (3.4.1).

Definitions 3.4.1:

(1) $A \subset R^n$ is said to be *self-invariant* for the differential system (3.4.1) if

$$x_0 \in A \Rightarrow x(t) \in A \quad t \geq t_0, t \in \mathbb{T};$$

(2) B is said to be conditionally-invariant with respect to A for the differential system (3.4.1) if

$$x_0 \in A \Rightarrow x(t) \in B, \ t \geq t_0, t \in \mathbb{T}.$$

Definitions 3.4.2: The conditionally invariant set B relative to A is said to be uniformly asymptotically stable (U.A.S.) if

(i) it is *uniformly stable* (U.S.), i.e., if given any $\epsilon > 0$ and $t_0 \in \mathbb{T}$, there is a $\delta = \delta(\epsilon) > 0$ such that

$$x_0 \in S(A, \delta) \text{ implies that } x(t) \in S(B, \epsilon), \ t \geq t_0$$

where

$$S(A, \delta) = \{x \in R^n \colon |\, x \,| \ \leq r \,|\, \delta\},$$

and

$$S(B, \epsilon) = \{x \in R^n \colon |\, x \,| \ \leq \rho + \epsilon\};$$

(ii) it is *uniformly quasi-asymptotically stable*, i.e., given $0 < \eta < \epsilon$, $t_0 \in R_+$, there exist a $\delta_0 = \delta_0(\eta)$, $0 < \delta_0 < \delta$, and $T = T(\eta)$ such that

$$|\, x_0 \,| \ < S(A, \delta_0) \text{ implies that } |\, x(t) \,| \ < S(B, \eta) \ t \geq t_0 + T.$$

We need corresponding definitions relative to the comparison equation (3.4.2). First we give the following definitions.

Let $\mathcal{K} = \{a \in C_{rd}[\mathbb{T}, R_+]: a(u)$ is strictly increasing in u, $a(0) = 0$ and $a(u) \to \infty$ as $u \to \infty\}$

Definitions 3.4.3: Let $\Omega: \{u \in R_+: u \leq r_0\}$, for some $r_0 > 0$.

(1) Ω is said to be *self-invariant* if

$$u_0 \leq r_0 \text{ implies that } u(t) \leq r_0, t \geq t_0, \ t \in \mathbb{T}.$$

(2) Ω is said to be *U.A.S.* if

(i) it is U.S., i.e., given any $\epsilon > 0$ and $t_0 \in \mathbb{T}$, there exists a $\delta = \delta(\epsilon) > 0$ such that $u_0 < r_0 + \delta)$implies $u(t) < r_0 + \epsilon, t \geq t_0$, and

(ii) given $\eta > 0$ and $t_0 \in R_+$, there exists a $\delta_0 > 0$ and $T = T(\eta) > 0$ such that $u_0 < r + \delta_0$ implies that $u(t) < r_0 + \eta, t \geq t_0 + T$.

We now prove the following theorem on U.A.S.

Theorem 3.4.1: *Assume that*

(A_0) *There exists* $V \in C_{rd}[\mathbb{T} \times R^n, R_+]$, $V(t, x)$ *locally Lipschitzian in* x *for each right-dense* $t \in \mathbb{T}$, *such that, for* $(t, x) \in \mathbb{T} \times R^n$,

$$b(\,|\,x\,|\,) \leq V(t,x) \leq a(\,|\,x\,|\,)$$

where $a, b \in \kappa$, *and*

$$D^+V^\Delta(t,x) \leq g(t, V(t,x), \,|\,\lambda\,|\,)$$

where $g \in C_{rd}[\mathbb{T} \times R_+ \times R_+, R]$;

(A_1)*For each* $\lambda \in R^d$, *there exists* $r = r(\lambda) > 0$, $\rho = \rho(\lambda) > 0$, *such that* $a(r) = b(\rho)$, *and* $r(\lambda)$, $\rho(\lambda) \to 0$ *as* $|\,\lambda\,| \to 0$;

(A_2)*The set* $\Omega = \{u \in R_+ : 0 \leq u \leq a(r)\}$ *is invariant and U.A.S.*

Then, the set $B = \{x \in R^m : \,|\,x\,|\, \leq \rho\}$ *is conditionally invariant with respect to* $A = \{x \in R^n : \,|\,x\,|\, \leq r\}$ *and it is U.A.S.*

Proof: Suppose that the set Ω is invariant with respect to (3.4.2) and U.A.S. We shall first show that the set $B = \{x \in R^m : \,|\,x\,|\, \leq \rho\}$ is conditionally invariant with respect to $A = \{x \in R^n : \,|\,x\,|\, \leq r\}$ relative to (3.4.1).

Since Ω is invariant relative to (3.4.2), we have $u_0 \leq a(r)$ implies that

$$u(t) \leq a(r), \quad t \geq t_0, \tag{3.4.4}$$

where $u(t)$ is any solution of (3.4.2). We need to prove that

$$|\,x_0\,| \leq r \text{ implies that } |\,x(t)\,| \leq \rho, t \geq t_0, \ t \in \mathbb{T}.$$

If not, there would exist a solution $x(t)$ of (3.4.1) with $|x_0| \leq r$ and $t_2 > t_0$, $t_2 \in \mathbb{T}$, such that

$$|x_0| \leq r \text{ but } |x(t_2)| > \rho \text{ and } |x(t)| \leq \rho,\ t_0 \leq t < t_2.$$

By Theorem we have,

$$V(t, x(t)) \leq w(t), \ t_0 \leq t \leq t_2$$

where $w(t) = w(t, |\lambda|, t_0, u_0)$ is the maximal solution of (3.4.2). Take $u_0 = V(t_0, x_0)$ so that when $|x_0| \leq r$, we have

$$u_0 = V(t_0, x_0) \leq a(|x_0|) \leq a(r).$$

As a result, it follows, in view of (3.4.4) and (A_1), that

$$b(\rho) < b(|x(t_2)|) \leq V(t_2, x(t_2)) \leq w(t_2) \leq a(r) = b(\rho),$$

which is a contradiction. Hence, B is conditionally invariant with respect to A.

Next, we need to prove U.S. of the conditionally invariant set B with respect to A. Let $\epsilon > 0$ and $t_0 \in \mathbb{T}$ be given. Choose $\delta_1 = \delta_1(\epsilon) > 0$ such that

$$a(r + \delta_1) < b(\rho + \epsilon).$$

Since the invariant set Ω is U.S., we have

$$u_0 < a(r + \delta) \Rightarrow u(t) < a(r + \delta_1) \text{ for } t \geq t_0, t \in \mathbb{T}.$$

Claim: $|x_0| < r+\delta \Rightarrow |x(t)| < \rho+\epsilon,\ t \geq t_0$. If this is not true, then there exists a solution of (3.4.1) and $t_1 > t_0$ such that

$$|x(t_1)| \geq \rho+\epsilon \text{ and } |x(t)| < \rho+\epsilon \ \ t_0 \leq t < t_1.$$

Choose $u_0 = V(t_0, x_0)$. Then, as before, by Theorem 3.1.1 we have, since

$u_0 = V(t_0, x_0) \leq a(|x_0|) < a(r+\delta)$,

$$b(\rho+\epsilon) \leq b(|x(t_1)|) \leq V(t, x(t_1)) \leq w(t_1) < a(r+\delta_1) < b(\rho+\epsilon).$$

This contradiction proves U.S. of Ω.

Finally, to prove U.A.S., fix $\epsilon = \rho_0$ and designate $\delta_0^0 = \delta(\rho_0)$.Since Ω is U.A.S., given $0 < \eta < \rho_0$ and $t_0 \in \mathbb{T}$, there exists a $\delta_0^* > 0$ and $T = T(\eta) > 0$ such that

$$u_0 < a(r+\delta_0^*) \text{ implies that } u(t) < a(r+\eta), \quad t \geq t_0 + T.$$

Choose $\delta_0 = min(\delta_0^0, \delta_0^*)$ so that we have, by U.S. of B,

$$|x_0| < r+\delta_0 \text{ implies that } |x(t)| < \rho+\rho_0,\ t \geq t_0.$$

Consequently, with $u_0 \equiv V(t_0, x_0) \leq a(r+\delta)$, we arrive at, for $t \geq t_0 + T$,

$$b(|x(t)|) \leq V(t, x(t)) \leq r(t) < a(r+\eta) < b(\rho+\eta)$$

which implies that $|x(t)| < \rho+\eta,\ t \geq t_0 + T$, provided $|x_0| < r+\delta_0$. The proof is therefore complete.

As an application of Theorem 3.4.1, we shall study the control of uncertain systems. Consider the dynamical system

$$x^{\Delta} = f(t,x,\omega) + B(t,x)F(t,x,u,\omega) \tag{3.4.5}$$

$$x(t_0) = x_0, \quad t \in \mathbb{T},$$

where $u \in R^m$ is the control and $\omega \in \emptyset$ is an uncertain parameter. Let us list the following conditions for convenience.

(i) $f \in C_{rd}[\mathbb{T} \times R^n \times \Omega, R^n]$, $B \in C_{rd}[\mathbb{T} \times R^n, R^{n \times m}]$ is a matrix and $F \in C_{rd}[\mathbb{T} \times R^n \times R^m \times \Omega, R^m]$.

(ii) $V \in C'_{rd}[\mathbb{T} \times R^n, R_+]$, and for $(t,x) \in \mathbb{T} \times R^n$

$$V_f^{\Delta}(t,x) \leq -c(|x|)$$

$$b(|x|) \leq V(t,x) \leq a(|x|),$$

where $a, b, c \in \mathcal{K}$.

(iii) For each $\omega \in \emptyset$

$$u^T F(t,x,u,\omega) \geq -\beta_1(t,x,\omega)|u| + \beta_2(t,x,\omega)|u|^2$$

where $\beta_1, \beta_2 \in C_{rd}[\mathbb{T} \times R^n \times \phi, R_+]$ satisfying

$$\beta_1(t,x,\omega) \leq \beta_2(t,x,\omega)\rho(t,x),$$

$$\beta_1(t,x,\omega) \leq k(t,x),$$

$\rho, k \in C_{rd}[\mathbb{T} \times R^n, R_+]$.

(iv) $p = \{p_\lambda \in C_{rd}[\mathbb{T} \times R^n, R^n] : \lambda > 0\}$ is the stabilizing family of controllers satisfying

$$| \alpha(t,x) | \, p_\lambda(t,x) = - \, | p_\lambda(t,x) | \, \alpha(t,x)$$

(i.e., p_λ is of opposite direction to α), and

$$| \eta | > 0 \Rightarrow | p_\lambda | \geq \rho[1 - \tfrac{\lambda}{|\eta|}]$$

where $\alpha = B^T V_x^T$ and $\eta(t,x) = k(t,x) \cdot \alpha(t,x)$.

We can now prove the following result as an application of Theorem 3.4.1. Let A and B be the sets defined in 3.4.3.

Theorem 3.4.2: *Suppose that the conditions (i)-(iv) hold. Then the set B is conditionally invariant with respect to the set and U.A.S.*

Proof: From (ii), it follows that $V_f^\Delta(t,x) \leq -\gamma(v(t,x))$ where $\gamma(u) = c(a^{-1}(u))$. Using conditions (iii) and (iv), we obtain

$$\alpha(t,x)F(t,x,p_\lambda,\omega) = - \tfrac{|\alpha|}{|p_\lambda|} p_\lambda F(t,x,p_\lambda,\omega)$$

$$\leq | \alpha | \, [\beta_1(t,x) - \beta_2(t,x) \, | p_\lambda |]$$

$$\leq | \alpha | \, [\beta_1(t,x) - \beta_2(t,x)\rho(t,x)(1 - \tfrac{\lambda}{|\eta|})$$

$$\leq \tfrac{|\alpha|\beta_1(t,x)}{|\eta|}\lambda \leq \lambda.$$

Hence, it follows that

$$V_f^{\Delta} \leq -\gamma(V(t,x)) + \lambda$$

and consequently, $g(t,w,\lambda) = -\gamma(w) + \lambda$. Clearly, $a(r) = a(r(\lambda)) = \gamma^{-1}(\lambda)$, and because of (ii), we see that (A_0) and (A_1) of Theorem 3.4.1 are satisfied. It is therefore enough to show that (A_2) is true relative to (3.4.2).

To prove that $0 \leq u_0 \leq a(r) \Rightarrow u(t) \leq a(r)$ for $t \geq t_0$, $t \in \mathbb{T}$, suppose that it is note true. Then, there exists $t^* > t_0$ such that $u(t^*) > a(r)$ and $t_1 \in (t_0, t^*)$ such that $u(t_1) \leq a(r)$ and $u(t) \geq a(r)$ for $t_1 \leq t \leq t^*$. Hence,

$$a(r) < u(t^*) = u(t_1) + \int_{t_1}^{t^*} g(s, u(s), \lambda) ds \leq a(r)$$

since $u(t_1) \leq a(r)$ and $g(t, u, |\lambda|) \leq 0$ if $u \geq a(r)$. This contradiction proves that the set $\Omega = \{u \in R_+ : 0 \leq u \leq a(r)\}$ is self-invariant.

To prove U.S. of Ω, let $\epsilon > 0$, $t_0 \in \mathbb{T}$ be given. Choose $\delta = \delta(\epsilon)$ such that $a(r+\delta) < a(r+\epsilon)$. Then, we claim

$$u_0 < a(r+\delta) \Rightarrow u(t) < a(r+\epsilon), t \in \mathbb{T}$$

If not, there exists $t_1 \geq t_0$ such that $u_0 < a(r+\delta)$ and $u(t_1) \geq a(r+\epsilon)$. This, in turn, yields

$$a(r+\epsilon) \leq u(t_1) < a(r+\delta) < a(r+\epsilon).$$

This contradiction proves U.S. of Ω.

To show that Ω is U.A.S., set $\rho_0 \equiv \epsilon$ and $\delta_0 = \delta(\rho_0)$, so that by U.S. of Ω

$$u_0 < a(r+\delta_0) \Rightarrow u(t) < b(\rho+\rho_0),\ t \in \mathbb{T}$$

Assume $u_0 < a(r+\delta_0)$. Let $0 < \eta < \rho_0$ and $\delta = \delta(\eta)$, $0 < \delta = \delta(\eta) < \delta_0$, of U.S.

Let $T > \frac{a(r+\delta_0)}{\gamma(a(r+\delta(\eta))-\lambda}$. Then, we claim that, for some $t^* \in [t_0, t_0+T]$ we have $u(t^*) < a(r+\delta)$. If this is false then we have $u(t) \geq a(r+\delta)$ for $t \in [t_0, t_0+T]$, which yields

$$0 \leq a(r+\delta(\eta)) \leq u(t_0+T) = \mu(t_0) + \int_{t_0}^{t_0+T} (-\gamma(u(s)) + \lambda)ds$$

$$\leq a(r+\delta_0) + [-\gamma(a(r+\delta(\eta)) + \lambda]T.$$

In view of the choice of T, this leads to a contradiction. Hence, it follows that, if $u_0 < a(r+\delta_0)$, then $u(t) < \eta$ for $t \geq t_0 + T$, which implies that Ω is U.A.S. and the proof of the theorem is complete.

We shall, next, consider results that are substantial refinements of the foregoing theorems. For this purpose, we need another comparison result which is dual to Theorem 3.1.1, which we merely state since its proof is similar to the proof of Theorem 3.1.1.

Theorem 3.4.3: *Let $V \in C_{rd}[\mathbb{T} \times R^n, R_n]$, $V(t,x)$ be locally Lipschitzian in x for each $t \in \mathbb{T}$ which is rd, and let*

$$D^+V^\Delta(t,x) \geq g(t, V(t,x))$$

where $g \in C_{rd}[\mathbb{T} \times R_+, R]$, $g(t,u)\,\mu^*(t) + u$ is nondecreasing in u for each $t \in \mathbb{T}$ and $\rho(t) = \rho(t, t_0, u_0)$ is the minimal solutoin of $u^\Delta = g(t,u)$, $u(t_0) = u_0 \geq 0$, existing on $\mathbb{T}$. Then, $V(t_0, x_0) \geq u_0$ implies that $V(t, x(t)) \geq \rho(t)$, $t \in \mathbb{T}$, $t \geq t_0$.

We need both comparision results in our discussion below.

Consider the dynamic system on time scale

$$x^\Delta = f(t, x, \lambda), \;\; x(t_0) = x_0, \; t_0 \in \mathbb{T} \tag{3.4.6}$$

where $f \in C_{rd}[\mathbb{T} \times R^n \times R^d, R^n], \lambda \in R^d$ is an uncertain parameter and $\mathbb{T}$ is the time scale. Consider also the comparison dynamic equation

$$u^\Delta = g(t, u, \mu), \; u(t_0) = u_0 \geq 0, \tag{3.4.7}$$

where $g \in C_{rd}[\mathbb{T} \times R_+^2, R]$ and $\mu = \mu(\lambda) \geq 0$ is a parameter depending on λ.

Let $\rho_0 \leq r_0 \leq r \leq \rho$ depending on λ. Then we shall say that the set $B = [x \in R^n : \rho_0 \leq |x| \leq \rho]$ is conditionally invariant with respect to $A = [x \in R^n : r_0 \leq |x| \leq r]$ and is uniformly asymptotically stable (UAS) relative to (3.4.6) if

(i) $r_0 \leq |x_0| \leq r$ implies $\rho_0 \leq |x(t)| \leq \rho, t \in \mathbb{T}, t \geq t_0$;

(ii) given $\epsilon > 0$ and $t_0 \in \mathbb{T}$,

  (a) there exists a $\delta = \delta(\epsilon) > 0$ such that $r_0 - \delta \leq |x_0| \leq r + \delta$ implies $\rho_0 - \epsilon \leq |x(t)| \leq \rho + \epsilon$, $t \geq t_0, t \in \mathbb{T}$;

  (b) there exists a $\delta_0 > 0$ and a $T = T(\epsilon) > 0$ such that $r_0 - \delta_0 \leq |x_0| \leq r + \delta_0$ implies $\rho_0 - \epsilon \leq |x(t)| \leq \rho + \epsilon$, $t \geq t_0 + T$, $t \in \mathbb{T}$;

  where $x(t) = x(t, t_0, x_0)$ is any solution of (3.4.6).

Relative to the comparison equation (3.4.7), we shall say that $\Omega = [u \in R_+ : R_0 \leq u \leq R]$ is invariant and is UAS relative to (3.4.7) if

(i) $R_0 \leq u_0 \leq R$ implies $R_0 \leq u(t) \leq R, t \geq t_0, t \in \mathbb{T}$;

(ii) given $\epsilon > 0$ and $t_0 \in \mathbb{T}$,

  (a) there exists a $\delta = \delta(\epsilon) > 0$ such that $R_0 - \delta \leq u_0 \leq R + \delta$ implies $R_0 - \epsilon \leq u(t) \leq R + \epsilon$, $t \geq t_0, t \in \mathbb{T}$;

  (b) there exists a $\delta_0 > 0$ and a $T = T(\epsilon) > 0$ such that $R_0 - \delta_0 \leq u_0 \leq R + \delta_0$ implies $R_0 - \epsilon \leq u(t) \leq R + \epsilon$, $t \geq t_0 + T$, $t \in \mathbb{T}$;

  where $u(t) = u(t, t_0, u_0)$ is any solution of (3.4.7).

We can now prove the following result on UAS of the conditionally invariant set B with respect to A, relative to the system (3.4.6). Let us define the sets Ω_r, Ω_{r_0} by $\Omega_r = [x \in R^n : x \in A\,\mu^*(t)$ and $|x| \geq r], \Omega_{r_0} = [x \in R^n : x \in A\,\mu^*(t)$ and $|x| \leq r_0]$.

Theorem 3.4.4: *Assume that*

(A_0)*for each* $\lambda \in R^d$, *there exist* $r = r(\lambda), r_0 = r_0(\lambda), r_0 \leq r$ *satisfying* $r \to 0$ *as* $|\lambda| \to 0$ *and* $r_0 \to \infty$ *as* $|\lambda| \to \infty$;

(A_1)*there exists* $V \in C_{rd}[\mathbb{T} \times R^n, R_+]$, $V(t,x)$ *be locally Lipschitzian in* x *for each right dense* $t \in \mathbb{T}$ *and for* $a_i, b_i \in K$, $i = 1, 2$,

$$b_1(|x|) \leq V(t,x) \leq a_1(|x|) \text{ if } x \in \Omega_r$$

$$b_2(|x|) \leq V(t,x) \leq a_2(|x|) \text{ if } x \in \Omega_{r_0}$$

(A_2)*if* $x \in \Omega_r$, $D^+V^\Delta(t,x) \leq g(t, V(t,x), r)$, *and if* $x \in \Omega_{r_0}$, $D^+V^\Delta(t,x) \geq g(t, V(t,x), r_0)$, *where* $g \in C_{rd}[\mathbb{T} \times R_+^2, R]$, $g(t,u,\mu)\,\mu^*(t) + u$ *is nondecreasing in* u *for each* (t,u);

(A_3)*for each* $r_0 \leq r$, *there exists* $R_0 \leq R$ *such that* $R = a_1(r) = b_1(\rho)$ *and* $R_0 = b_2(r_0) = a_2(\rho_0)$, *where* $\rho_0 \leq r_0 \leq r \leq \rho$ *and* $R \to 0$ *as* $r \to 0$, $R_0 \to \infty$ *as* $r_0 \to \infty$;

(A_4)*the set* Ω *is invariant and is UAS with respect to* (3.4.7).

Then the set B *is conditionally invariant with respect to* A *and is UAS relative to the system* (3.4.6).

Proof: We shall first prove that B is conditionally invariant with respect to A and (3.4.6). If not, there would exist a solution $x(t) = x(t, t_0, x_0)$ of (3.4.1) with $r_0 \leq |x_0| \leq r$ and $t_0 < t_2$ such that either

(i) $|x(t_2)| > \rho$ and $r_0 \leq |x(t)|$, $t \in [t_0, t_2] \cap \mathbb{T}$,

(ii) $|x(t_2)| < \rho_0$ and $|x(t)| \leq r$, $t \in [t_0, t_2] \cap \mathbb{T}$.

Because of (A_2), using comparison Theorem 3.1.1, 3.4.3 we get either

$$V(t, x(t)) \leq r(t, t_0, V(t_0, x_0)),$$

or $$V(t, x(t)) \geq \rho(t, t_0, V(t_0, x_0)),$$

for $t \in [t_0, t_2] \cap \mathbb{T}$, where $r(t, t_0, u_0)$, $\rho(t, t_0, u_0)$ are the maximal and minimal solutions of (3.4.7). Hence using (A_3) and (A_4). in case (i), we have

$$b_1(\rho) < b_1(|x(t_2)|) \leq V(t_2, x(t_2)) \leq r(t_2, t_0, V(t_0, x_0))$$

$$\leq r(t_2, t_0, a_1(|x_0|)) \leq r(t_2, t_0, a_1(r)) \leq a_1(r) = b_1(\rho),$$

or, in case (ii), we get

$$a_2(\rho_0) > a_2(|x(t_2)|) \geq V(t_2, x(t_2)) \geq \rho(t_2, t_0, V(t_0, x_0))$$

$$\geq \rho(t_2, t_0, b_2(|x_0|)) \geq \rho(t_2, t_0, b_2(r_0)) \geq b_2(r_0) = a_2(\rho_0).$$

Thus, we have a contradiction in both cases and hence B is conditionally invariant with respect to A and (3.4.6).

Let $0 < \epsilon < \rho_0$ and $t_0 \in \mathbb{T}$ be given. Since (A_4) holds and $a_1(r) = b_1(\rho) = R$, $R_0 = b_2(r_0) = a_2(\rho_0)$, given $a_2(\rho_0 - \epsilon)$, $b_1(\rho + \epsilon)$, there exist $\epsilon_1, \delta_1, \delta > 0$ such that

$$R + \delta_1 = a_1(r + \delta) < b_1(\rho + \epsilon) = R + \epsilon_1,$$

and $\quad R_0 - \epsilon = a_2(\rho_0 - \epsilon) < b_2(r_0 - \delta) = R_0 - \delta_1,$

satisfying

$R_0 - \delta_1 < u_0 < R + \delta_1$ implies $R_0 - \epsilon_1 < u(t) < R + \epsilon_1, t \geq t_0, t \in \mathbb{T},$

where $u(t) = u(t, t_0, u_0)$ is any solution of (3.4.7). We claim that with this $\delta > 0$, the set B is US relative to A, that is,

$$r_0 - \delta < |x_0| < r + \delta \text{ implies } \rho_0 - \epsilon < |x(t)| < \rho + \epsilon,\ t \geq t_0, t \in \mathbb{T}.$$

If this is not true, there would exist a solution $x(t)$ of (3.4.6) with $r_0 - \delta < |x_0| < r + \delta$ and a $t_2 > t_0$ such that either

$$(a)\ \ |x(t_2)| \geq \rho + \epsilon \text{ and } |x(t)| \geq r_0,\ [t_0, t_2] \cap \mathbb{T},$$

or $\quad (b)\ \ |x(t_2)| \leq \rho_0 - \epsilon$ and $|x(t)| \leq r,\ [t_0, t_2] \cap \mathbb{T}.$

Consider (a). As before, we obtain

$$V(t, x(t)) \leq r(t, t_0, V(t_0, x_0)),\ [t_0, t_2] \cap \mathbb{T},$$

and therefore, we arrive at the contradiction

$$b_1(\rho + \epsilon) \leq b_1(|x(t_2)|) \leq V(t_2, x(t_2)) \leq r(t_2, t_0, V(t_0, x_0))$$

$$\leq r(t_2, t_0, a_1(|x_0|)) \leq r(t_2, t_0, a_1(r + \delta)) < b_1(\rho + \epsilon).$$

Similarly, in case (b), we first get,

$$V(t, x(t)) \geq \rho\,(t, t_0, V(t_0, x_0)), \ \ [t_0, t_2] \cap \mathbb{T},$$

and then it follows that

$$a_2(\rho_0 - \epsilon\,) > a_2(|x(t_2)|) \geq V(t_2, x(t_2)) \geq \rho(t_2, t_0, V(t_0, x_0))$$

$$\geq \rho(t_2, t_0, b_2(|x_0|)) \geq \rho(t_2, t_0, b_2(r_0 - \delta)) \geq b_2(r_0) = a_2(\rho_0 - \epsilon),$$

which is again a contradiction. Hence the set B is US relative to A.

To prove UAS of the set B relative to A, let us fix $\epsilon = \rho_0$ and designate by $\delta_0 = \delta(\rho_0)$ so that we have

$$r_0 - \delta_0 < |x_0| < r + \delta_0 \text{ implies } 0 < |x(t)| < \rho + \rho_0,\ t \geq t_0, t \in \mathbb{T}.$$

Let $0 < \epsilon < \rho_0$ and $t_0 \in \mathbb{T}$. Since Ω is UAS, given $a_2(\rho_0 - \epsilon)$, $b_1(\rho + \epsilon)$, there exists a $T = T(\epsilon) > 0$, with $t_0 + T \in \mathbb{T}$ such that $b_2(r_0 - \delta_0) < u_0 < a_1(r + \delta_0)$ implies $a_2(\rho_0 - \epsilon\,) < u(t) < b_1(\rho + \epsilon)$, $t \geq t_0 + \mathbb{T}$.

We claim that whenever $r_0 - \delta_0 < |x_0| < r + \delta_0$, we have

$$\rho_0 - \epsilon\, < |x(t)| < \rho + \epsilon,\ t \geq t_0 + \mathbb{T}, t \in \mathbb{T}.$$

If this is not true, there would exist a solution $x(t)$ of (3.4.6) such that

(a) $|x(t_2)| \geq \rho + \epsilon\,,\ t_2 \geq t_0 + \mathbb{T},\ t_2 \in \mathbb{T}$,

(b) $|x(t_2)| \leq \rho_0 - \epsilon\,, t_2 \geq t_0 + \mathbb{T},\ t_2 \in \mathbb{T}$,

where $r_0 - \delta_0 < |x_0| < r + \delta_0$. As before, using (A_2) and (A_3), we get successively

$$b_1(\,\rho + \epsilon\,) \leq V(t_2, x(t_2)) \leq r(t_2, t_0, a_1(r + \delta_0)) < b_1(\,\rho + \epsilon\,),$$

and $a_2(\rho_0 - \epsilon) \geq V(t_2, x(t_2)) \geq \rho(t_2, t_0, b_2(r_0 - \delta_0)) > a_2(\rho_0 - \epsilon)$,

which are contradictions. Hence we have B is UAS with respect to A relative to the system (3.4.6) and the proof is complete.

Remarks: If $\mathbb{T} = R$, then (3.4.6), (3.4.7) reduce to the continuous differential systems. Since, in this case, $\mu^*(t) = 0$, the results of Theorem 3.4.4 reduce to those in [47]. Note that the conditions (A_1) and (A_2) are then weaker which are sufficient to prove UAS. If, on the other hand, $\mathbb{T} = Z$, so that $\mu^*(t) = 1$, (3.4.6) and (3.4.7) reduce to difference equations, and consequently, one needs stronger conditions (A_1), (A_2). Theorem 3.4.4 offers new results in this special case.

As an application of Theorem 3.4.4, we shall consider the control of uncertain dynamic system on time scales of the form

$$x^\Delta = f_0(t, x, \lambda) + B(t, x, \lambda)\ F(t, x, u, \lambda)\ \ x(t_0) = x_0,\ t_0 \in \mathbb{T}, \quad (3.4.8)$$

under the following assumptions

(B_0) $f_0 \in C_{rd}[\mathbb{T} \times R^n \times \Omega_0, R^n]$, $B \in C_{rd}[\mathbb{T} \times R^n \times \Omega_0, R^{n \times m}]$, $F \in C_{rd}[\mathbb{T} \times R^n \times R^m \times \Omega, R^m]$, $\Omega_0 \subset R^d$ is a nonempty set and $u \in R^m$ is the control function.

(B_1) there exist $r_0 = r_0(\lambda) \leq r = r(\lambda)$ and $V \in C^1_{rd}[\mathbb{T} \times R^n, R_+]$ such that $V^\Delta_{f_0}(t, x) \leq -c_1(V(t, x))$ if $x \in \Omega_r$, $V^\Delta_{f_0}(t, x) \geq -c_2(V(t, x))$

if $x \in \Omega_{r_0}$, where $V_{f_0}^{\Delta}(t,x) = V_t^{\Delta}(t,x) + V_x(\sigma(t),x)\, f_0(t,x,\lambda)$, $c_i \in K$, $i = 1,2$ and $c_i(u)\mu^*(t) + u$ is nondecreasing in u;

(B_2) $b_1(|x|) \le V(t,x) \le a_1(|x|)$ if $x \in \Omega_r$, $b_2(|x|) \le V(t,x) \le a_2(|x|)$ if $x \in \Omega_{r_0}$ where $a_i, b_i \in K$, $i = 1,2$;

(B_3) for $x \in \Omega_r$, $u^T F(t,x,u,\lambda) \ge -\beta_1(t,x)|u| + \beta_2(t,x)|u|^2$ where $\beta_1, \beta_2 \in C_{rd}[\mathbb{T} \times R^n, R_+]$, $\beta_1 \le \beta_2 \widetilde{\rho}$, $\beta_1 \le k$, $\widetilde{\rho}, k \in C_{rd}[\mathbb{T} \times R^n, R_+]$;

(B_4) for $x \in \Omega_{r_0}$, $u^T F(t,x,u,\lambda) \le -r_1(t,x)|u| + r_2(t,x)|u|^2$ where $r_1, r_2 \in C_{rd}[\mathbb{T} \times R^n, R_+]$, $r_1 \ge r_2 \widetilde{\rho}$, $r_1 \ge k$;

(B_5) $P = [p_\mu \in C_{rd}[\mathbb{T} \times R^n, R^m]$ for $\mu > 0$ is the stabilizing family of controllers satisfying $|\alpha|\, p_\mu = -|p_\mu|\, \alpha$, $\alpha = B^T V_x^T$ and $\eta = k\,\alpha$, and if $|\eta| > 0$, $x \in \Omega_r$, $|p_\mu| \ge \widetilde{\rho}\,(1 - \frac{r}{|\eta|})$, if $|\eta| > 0$, $x \in \Omega_{r_0}$, $|p_\mu| \le \widetilde{\rho}\,(1 - \frac{r_0}{|\eta|})$.

We are now in a position to prove the following result.

Theorem 3.4.5: *Assume that conditions* (B_0) *to* (B_5) *hold. Suppose further that* $c_2^{-1}(u) \le c_1^{-1}(u)$. *Then the set* B *is conditionally invariant with respect to* A *and is UAS relative to the system* (3.4.8).

Proof: Let us first consider the case $x \in \Omega_r$. Then we have

$$V^{\Delta}_{f_0}(t,x) \leq -c_1(V(t,x))$$

and
$$\alpha(t,x,\lambda)\,F(t,x,p_\mu,\lambda) = -\frac{|\alpha(t,x,\lambda)|}{|p_\mu(t,x)|}\,F(t,x,p_\mu,\lambda)\,p_\mu$$

$$\leq |\alpha(t,x,\lambda)|\,[\,\beta_1(t,x) - \beta_2(t,x)\,|p_\mu(t,x)|\,]$$

$$\leq |\alpha(t,x,\lambda)|\,[\,\beta_1(t,x) - \beta_2(t,x)\,\widetilde{\rho}\,(t,x)\,(1 - \tfrac{r}{|\eta|})\,]$$

$$\leq |\alpha(t,x,\lambda)|\,\beta_1(t,x)\tfrac{r}{|\eta|} \leq r,$$

and therefore, we obtain

$$V^{\Delta}_{(3.4.3)}(t,x) \leq -c_1(V(t,x)) + r \quad \text{if } x \in \Omega_r$$

Similarly, if if $x \in \Omega_{r_0}$, we get

$$V^{\Delta}_{f_0}(t,x) \geq -c_2(V(t,x)),$$

and
$$\alpha(t,x,\lambda)\,F(t,x,p_\mu,\lambda) = -\frac{|\alpha(t,x,\lambda)|}{|p_\mu(t,x)|}\,F(t,x,p_\mu,\lambda)\,p_\mu$$

$$\geq |\alpha(t,x,\lambda)|\,[\,r_1(t,x) - r_2(t,x)\,|p_\mu(t,x)|\,]$$

$$\geq |\alpha(t,x,\lambda)|\,[\,r_1(t,x) - r_2(t,x)\,\widetilde{\rho}\,(t,x)\,(1 - \tfrac{r_0}{|\eta|})\,]$$

$$\geq |\alpha(t,x,\lambda)|\,r_1(t,x)\tfrac{r_0}{|\eta|} \geq r_0,$$

Thus we have

$$V^{\Delta}_{(3.4.3)}(t,x) \geq -c_2(V(t,x)) + r_0 \quad \text{if } x \in \Omega_{r_0}$$

This implies that

$$g(t,u,r) = -c_1(u) + r, \quad g(t,u,r_0) = -c_2(u) + r_0$$

and therefore $u = c_1^{-1}(u) = R$, $u = c_2^{-1}(u) = R_0$. Hence, in view of the properties of a_i, b_i, c_i, $i = 1,2$, we can find, for each λ, $\rho_0 \leq r_0 \leq r \leq \rho$ such that $R = a_1(r) = b_1(\rho)$, $R_0 = b_2(r_0) = a_2(\rho_0)$ and $R_0 \leq R$.

To apply Theorem 3.4.4, we need to show that the set $\Omega = [u \in R_+ : R_0 \leq u \leq R]$ is invariant and is UAS relative to the comparison dynamic equation (3.4.7). Because of the specific nature of g, it is not difficult to prove it following the proof of Theorem 3.4.4. We omit the details to avoid monotony. The proof of Theorem 3.4.5 is therefore complete.

3.5 Stability in Terms of Two Measures

The concepts of Lyapunov Stability have given rise to many new notions that are important in applications. For example, partial stability, conditional stability, eventual stability, practical stability and corresponding notions relative to boundedness to name a few. A notion which unifies and includes several known concepts of stability in a simple set up , is the notion of stability in terms of two different measures. This well developed for differential equations. In this section, we shall present some typical results in the framework of dynamic systems on time scales. Before proceeding further, we need some class of functions and concepts.

Let us first define the following classes of functions. We already have the class $\mathcal{K}$. We need others.

$$\mathcal{L} = \{\sigma \in C[R_+, R_+]: \sigma(u) \text{ is strictly decreasing in } u \text{ and } \lim_{u \to \infty} \sigma(u) = 0\},$$

$$C\mathcal{K} = \{a \in C_{rd}[\mathbb{T} \times R_+, R_+]: a(t,s) \in \mathcal{K} \text{ for each } t\},$$

$$\Gamma = \{h \in C_{rd}[\mathbb{T} \times R^n, R_+]: \inf_{(t,x)} \ h(t,x) = 0\},$$

We now define stability concepts for the system (3.1.1) in terms of two measures $h_0, h \in \Gamma$.

Definition 3.5.1: The dynamic system (3.1.1) is said to be

(S_1) (h_0, h)-*equistable* if, for each $\epsilon > 0$, $t_0 \in \mathbb{T}$, there exists a positive function $\delta = \delta(t_0, \epsilon)$ that is rd-continuous in t_0 for each ϵ such that $h_0(t_0, x_0) < \delta$ implies $h(t, x(t)) < \epsilon$, $t \geq t_0$, where $x(t) = x(t, t_0, x_0)$ is any solution of system (3.1.1),

(S_2) (h_0, h)-*uniformly stable* if the δ in (S_1) is independent of t_0.

Other stability concepts for the system (3.1.1) may be defined similarly based on the definitions given in Section 3.2.

Definition 3.5.2: Let $h_0, h \in \Gamma$. Then we say that

(i) h_0 is *finer* than h if there exists a $\rho > 0$ and a function $\varphi \in C\mathcal{K}$ such that $h_0(t,x) < \rho$ implies $h(t,x) \leq \varphi(t, h_0(t,x))$;

(ii) h_0 is *uniformly finer* than h if in (i) φ is independent of t.

Definition 3.5.3: Let $V \in C_{rd}[\mathbb{T} \times R^n, R_+]$. Then V is said to be

(i) *h-positive definite* if there exists a $\rho > 0$ and a function $b \in \mathcal{K}$ such that $b(h(t,x)) \leq V(t,x)$ whenever $h(t,x) < \rho$;

(ii) *h-decrescent* if there exists a $\rho > 0$ and a function $a \in \mathcal{K}$ such that $V(t,x) \leq a(h(t,x))$ whenever $h(t,x) < \rho$;

(iii) *h-weakly decrescent* if there exists a $\rho > 0$ and a function $a \in C\mathcal{K}$ such that $V(t,x) \leq a(t, h(t,x))$ whenever $h(t,x) < \rho$.

We are now in a position to prove stability criteria in terms of two measures.

Theorem 3.5.1: *Assume that*

(i) $V \in C_{rd}[\mathbb{T} \times R^n, R_+]$, $h \in \Gamma$, $V(t,x)$ *is locally Lipschitzian in* x *for each* $t \in \mathbb{T}$ *which is* rd, *and h-positive definite;*

(ii) $D^+V^\Delta(t,x) \leq 0, (t,x) \in S(h,\rho)$, *where*

$$S(h,\rho) = \{(t,x) \in \mathbb{T} \times R^n : h(t,x) < \rho, \rho > 0\}.$$

Then

(A) *if, in addition,* $h_0 \in \Gamma$, h_0 *is finer than* h *and* $V(t,x)$ *is* h_0*-weakly decrescent, then the system* (3.1.1) *is* (h_0, h)*-equistable,*

(B) *if, in addition,* $h_0 \in \Gamma$, h_0 *is uniformly finer than* h, *and* $V(t,x)$ *is* h_0*-decrescent, then the system* (3.1.1) *is* (h_0, h)*-uniformly stable.*

Proof: Let us first prove (A). Since $V(t,x)$ is h_0-weakly decrescent, then for $t_0 \in \mathbb{T}$, $x_0 \in R^n$, there exist a constant $\rho_0 > 0$ and a function $a \in C\mathcal{K}$ such that

$$V(t_0, x_0) \le a(t_0, h_0(t_0, x_0)), \text{ provided } h_0(t_0, x_0) < \rho_0. \tag{3.5.1}$$

The fact that $V(t,x)$ is h-positive definite implies that there exist a constant $\rho > 0$ and a function $b \in \mathcal{K}$ such that

$$b(h(t,x)) \le V(t,x), \text{ whenever } h(t,x) \le \rho. \tag{3.5.2}$$

Also, by the assumption that h_0 is finer than h, there exist a constant $\rho_0 > 0$ and a function $\varphi \in C\mathcal{K}$ such that

$$h(t_0, x_0) \le \varphi(t_0, h_0(t_0, x_0)), \text{ if } h_0(t_0, x_0) < \rho_0, \tag{3.5.3}$$

where ρ_0 is chosen so that $\varphi(t_0, \rho_0) < \rho$.

Let $\epsilon \in (0, \rho_0)$ and $t_0 \in \mathbb{T}$ be given. By the assumption on a, there exists a $\delta = \delta(t_0, \epsilon) > 0$ that is rd-continuous in t_0 such that

$$a(t_0, \delta) < b(\epsilon). \tag{3.5.4}$$

Then $h_0(t_0, x_0) < \delta$ implies, by (3.5.1)-(3.5.4), that $b(h(t_0, x_0)) \leq V(t_0, x_0) \leq a(t_0, h_0(t_0, x_0)) < b(\epsilon)$, which in turn yields that $h(t_0, x_0) < \epsilon$. We now claim that for every solution $x(t) = x(t, t_0, x_0)$ of (3.1.1) with $h_0(t_0, x_0) < \delta$

$$h(t, x(t)) < \epsilon, t \geq t_0. \tag{3.5.5}$$

If this is not true, then there would exist a $t_1 > t_0$ such that

$$h(t_1, x(t_1)) \geq \epsilon \text{ and } h(t, x(t)) < \epsilon, t \in [t_0, t_1), \tag{3.5.6}$$

for some solution $x(t) = x(t, t_0, x_0)$ of (3.1.1). Set $m(t) = V(t, x(t))$ for $t \in [t_0, t_1]$.

Then, by Theorem 3.1.1, with $g(t, u) \equiv 0$, we get

$$V(t, x(t)) \leq V(t_0, x_0) \ , t_0 \leq t \leq t_1,$$

which, in view of (3.5.1),(3.5.4),(3.5.6) gives

$$b(\epsilon) \leq b(h(t_1, x(t_1)) \leq V(t_1, x(t_1)) \leq V(t_0, x_0) < b(\epsilon),$$

which is a contradiction. Hence (3.5.5) is true and the system (3.1.1) is (h_0, h)-equistable.

To prove (B), note that if $V(t,x)$ is h_0-decrescent and h_0 is uniformly finer than h, then the functions a and φ in (3.5.1) and (3.5.3) are independent of t. Consequently, it is easily seen that the constant δ can be chosen to be independent of t_0. Hence the system (3.1.1) is (h_0,h)-uniformly stable.

We next prove a result on (h_0,h)-uniform asymptotic stability.

Theorem 3.5.2: *Assume that*

(i) $h_0, h \in \Gamma$ *and* h_0 *is uniformly finer than* h,

(ii) $V \in C_{rd}[\mathbb{T} \times R^n, R_+], V(t,x)$ *is locally Lipschitzian in* x *for each right dense* $t \in \mathbb{T}$, *h-positive definite,* h_0*-decrescent and*

$$D^+V^\Delta(t,x) \leq -C(h_0(t,x)), (t,x) \in S(h,\rho), C \in \mathcal{K}.$$

Then the system (3.1.1) *is* (h_0,h)*-uniformly asymptotically stable.*

Proof: Since $V(t,x)$ is h-positive definite and h_0-decrescent, there exist constants $0 < \rho_0 \leq \rho$, and functions $a, b \in \mathcal{K}$ such that

$$b(h(t,x)) \leq V(t,x), (t,x) \in S(h,\rho) \tag{3.5.7}$$

and

$$V(t,x) \leq a(h_0(t,x)), \text{ if } h_0(t,x) < \rho_0. \tag{3.5.8}$$

It follows from Theorem 3.5.1 that the system (3.1.1) is (h_0, h)-uniformly stable. Thus we let $\epsilon = \rho_0$, and designate by $\delta_0 = \delta(\rho_0)$ so that we have

$$h_0(t_0, x_0) < \delta_0 \text{ implies } h(t, x(t)) < \rho_0, t \geq t_0,$$

where $x(t) = x(t, t_0, x_0)$ is any solution of (3.1.1).

Let $0 < \epsilon < \rho_0$ and $\delta = \delta(\epsilon)$ be the same δ as in Definition 3.1.1 for (h_0, h)-uniform stability. Assume that $h_0(t_0, x_0) < \delta_0$. Set $T = T(\epsilon) = \frac{a(\delta_0)}{C(\delta)} + 1$. To prove (h_0, h)-uniform asymptotic stability, it is enough to show that there exists a $t^* \in [t_0, t_0 + T]$, such that

$$h_0(t^*, x(t^*)) < \delta.$$

If this is not true, then there exists a solution $x(t) = x(t, t_0, x_0)$ of (3.1.1) with $h_0(t_0, x_0) < \delta_0$ such that

$$h_0(t, x(t)) \geq \delta, t \in [t_0, t_0 + T]. \tag{3.5.9}$$

Let $m(t) = V(t, x(t))$. Then proceeding as in the proof of corollary 3.2.2, we arrive at

$$V(t_0 + T, x(t_0 + T)) \leq V(t_0, x_0) \quad - \int_{t_0}^{t_0+T} C(h_0(s, x(s)))\Delta s,$$

which yields,

$$\int_{t_0}^{t_0+T} C(h_0(s, x(s)))\Delta s \leq V(t_0, x_0) \leq a(h_0(t_0, x_0)) \leq a(\delta_0).$$

On the other hand, from (3.5.9), together with the definition of $T(\epsilon)$ we obtain

$$\int_{t_0}^{t_0+T} C(h_0(s, x(s)))\Delta s \geq C(\delta)T > a(\delta_0),$$

which is a contradiction. Thus the proof of the theorem is complete.

Finally, we shall prove a result in the general set up of comparison principle.

Theorem 3.5.3: *Assume that*

$(A_0) h_0, h$ *are rd-continuous, belong to the class* Γ *and* h_0 *is uniformly finer than* h*;*

$(A_1) V \in C_{rd}[\mathbb{T} \times R^n, R_+]$, $V(t, x)$ *is locally Lipschitzian in* x *for each* $t \in \mathbb{T}$ *which is rd,* h*-positive definite and* h_0*-decrescent;*

$(A_2) g \in C_{rd}[\mathbb{T} \times R_+, R]$ *and* $g(t, 0) \equiv 0$*;*

$(A_3) D^+V^{\Delta}(t, x) \leq g(t, V(t, x)), (t, x) \in \mathbb{T} \times R^n$.

Then the stability properties of the trivial solution of (3.3.2) *imply the corresponding* (h_0, h)*-stability properties of* (3.1.1).

Proof: We shall only prove (h_0, h)-equiasymptotic stability of (3.1.1). For this purpose, let us first prove (h_0, h)-equistability.

Since V is h-positive definite, there exists a $b \in \mathcal{K}$ and $\rho > 0$ such that

$$b(h(t,x)) \leq V(t,x), \text{ whenever } h(t,x) < \rho. \tag{3.5.10}$$

Let $\epsilon > 0$ and $t_0 \in \mathbb{T}$ be given. Suppose that the trivial solution of (3.3.2) is equistable. Then given $b(\epsilon) > 0$ and $t_0 \in \mathbb{T}$, there exists a function $\delta_1 = \delta_1(t_0, \epsilon) > 0$ such that

$$u_0 < \delta_1 \text{ implies } u(t, t_0, u_0) < b(\epsilon), t \in \mathbb{T} \tag{3.5.11}$$

where $u(t, t_0, u_0)$ is any solution of (3.3.2). We choose $u_0 = V(t_0, x_0)$. Since V is h_0-decrescent and h_0 is uniformly finer than h, there exists a $\rho_0 > 0$ and functions ϕ, $a \in \mathcal{K}$ such that

$$V(t_0, x_0) \leq a(h_0(t_0, x_0)) \text{ if } h_0(t_0, x_0) < \rho_0, \text{and } h(t_0, x_0) \leq \phi(h_0(t_0, x_0)),$$

$$\text{if } h_0(t_0, x_0) < \rho_0, \text{ where } \rho_0 \text{ is chosen to satisfy } \phi(\rho_0) < \rho. \tag{3.5.12}$$

It then follows from (3.5.10) that

$$b(h(t_0, x_0)) \leq V(t_0, x_0) \leq a(h_0(t_0, x_0)), \text{if } h_0(t_0, x_0) < \rho_0 \tag{3.5.13}$$

Choose $\delta = \delta(t_0, \epsilon)$ such that $0 < \delta \leq \rho_0$, $a(\delta) < \delta_1$ and let $h_0(t_0, x_0) < \delta$. Then (3.5.12) shows that $h(t_0, x_0) < \epsilon$ since $\delta_1 < b(\epsilon)$. We claim that

$$h(t, x(t)) < \epsilon, t \geq t_0 \text{ whenever } h_0(t_0, x_0) < \delta,$$

where $x(t) = x(t, t_0, x_0)$ is any solution of (3.1.1) with $h_0(t_0, x_0) < \delta$. If this is not true, then there exists a $t_1 > t_0$ and a solution $x(t)$ of (3.1.1) such that

$$\epsilon \leq h(t_1, x(t_1)) \text{ and } h(t, x(t)) < \epsilon, t_0 \leq t \leq t_1, \tag{3.5.14}$$

in view of the fact that $h(t_0, x_0) < \epsilon$ whenever $h_0(t_0, x_0) < \delta$. Now by Theorem 3.1.1., we have

$$V(t, x(t)) \leq r(t, t_0, u_0), t_0 \leq t \leq t_1, \tag{3.5.15}$$

where $r(t, t_0, u_0)$ is the maximal solution of (3.3.2). Now the relations (3.5.10), (3.5.11), (3.5.14) and (3.5.15) yield

$$b(\epsilon) \leq V(t_1, x(t_1)) \leq r(t_1, t_0, u_0) < b(\epsilon),$$

a contradiction, proving (h_0, h)-stability of (3.1.1).

Suppose next that the trivial solution of (3.3.2) is equi-attractive. Then we have that, given $b(\epsilon) > 0$ from (h_0, h)-stability and $t_0 \in \mathbb{T}$, there exist positive numbers $\delta_1^* = \delta_1^*(t_0)$ and $T = T(t_0, \epsilon) > 0$ such that

$$u_0 < \delta_1^* \text{ implies } u(t, t_0, u_0) < b(\epsilon), t \geq t_0 + T. \tag{3.5.16}$$

Choosing $u_0 = V(t_0, x_0)$ as before, we find a $\delta_0^* = \delta_0^*(t_0) > 0$ such that $0 < \delta_0^* \leq \rho_0$ and $a(\delta_0^*) < \delta_1^*$. Let $h_0(t_0, x_0) < \delta_0$. Then the estimate (3.5.15) is valid for all $t \geq t_0$. Suppose now that there exists a sequence $\{t_k\}$, $t_k \geq t_0 + T, t_k \to \infty$ as $k \to \infty$ such that $\epsilon \leq h(t_k, x(t_k))$ where $x(t)$ is any solution of (3.1.1) such that $h_0(t_0, x_0) < \delta_0$. This leads to a contradiction

$$b(\epsilon) \leq V(t_k, x(t_k)) \leq r(t_k, t_0, u_0) < b(\epsilon)$$

because of (3.5.15) and (3.5.16). Hence the system (3.1.1) is (h_0, h)-equi-asymptotically stable and the proof is complete.

3.6 Vector Lyapunov Functions and Practical Stability

In this section, we shall discuss practical stability of dynamic system (3.1.1) in terms of two measures employing the method of Vector Lyapunov functions.

We need the following definitions.

Definition 3.6.1: The dynamic system (3.1.1) is said to be

(PS1) (h_0,h)-*practically stable* if, given λ, $A \in R_+$ with $0 < \lambda < A$, we have , $h_0(t_0,x_0) < \lambda$ implies $h(t, x(t)) < A$, $t \geq t_0$ for some $t_0 \in \mathbb{T}$, where $x(t) = x(t, t_0, x_0)$ is any solution of (3.1.1);

(PS2) (h_0,h)-*uniformly practically stable* if (PS1) holds for every $t_0 \in \mathbb{T}$;

(PS3) (h_0,h)-*practically quasi-stable* if, given $(\lambda,B,T) > 0$ and some $t_0 \in \mathbb{T}$, we have, $h_0(t_0, x_0) < \lambda$ implies $h(t, x(t)) < B$, $t \geq t_0+T$ where $x(t) = x(t, t_0, x_0)$ is any solution of (3.1.1);

(PS4) (h_0,h)-*uniformly practically quasi-stable* if (PS3) holds for all $t_0 \in \mathbb{T}$;

(PS5) (h_0,h)-*strongly practically stable* if (PS1) and (PS3) hold simultaneously;

(PS6) (h_0,h)-*strongly uniformly practically stable* if (PS1) and (PS3) hold simultaneously;

(PS7) (h_0,h)-*practically asymptotically stable* if (PS1) holds and $\lim_{t\to\infty} h(t,x(t)) = 0$.

Let us consider the following comparison system:

$$w^{\Delta} = G(t,w), \quad w(t_0) = w_0 \geq 0 \tag{3.6.1}$$

where $G \in C_{rd}[\mathbb{T} \times R^q, R^q]$. To state the first result we need the following definition:

Definition 3.6.2: The solution $w(t) = w(t,t_0,w_0)$ of (4.4.2) is said to be

(PS1)* *practically stable* with respect to (λ, A) if given $0 < \lambda < A$ $\sum_{i=1}^{q} w_{0_i} < \lambda$ implies $\sum_{i=1}^{q} w_i(t, t_0, w_0) < A, \quad t \geq t_0$ for some $t_0 \in \mathbb{T}\bigcap R_+$;

Based on this definition, one can formulate other notions of practical stability for comparison system (3.6.1).

We shall now consider practical stability criteria.

Theorem 3.6.1: *Assume that*

(A1) $0 < \lambda < A$

(A2) *h_0, $h \in \Gamma$, and h_0 is uniformly finer than h, i.e., $h(t,x) \leq \phi(h_0(t,x))$, $\phi \in \mathcal{K}$, whenever $h_0(t,x) < \lambda$;*

(A3) *there exists $V \in C_{rd}[\mathbb{T} \times R^n, R^p_+]$, and $Q \in C_{rd}[R^p_+, R^q_+]$ such that if $Q(V(t,x)) \equiv \mathcal{V}(t,x)$, $\mathcal{V}(t,u)$ is locally Lipschitzian in u and for $(t,x) \in S(h,A) = \{(t,x) \in \mathbb{T} \times R^n: h(t,x) < A\}$, $a, b \in \mathcal{K}$*

$$b(h(t,x)) \leq \sum_{i=0}^{q} \mathcal{V}_i(t,x) \qquad \text{if } h(t,x) < A$$

$$\sum_{0}^{\infty} \mathcal{V}_i(t,x) \leq a(h_0(t,x)) \qquad \text{if } h_0(t,x) < \lambda, \tag{3.6.2}$$

(A4) *for $(t,x) \in S(h,A)$*

$$D^+\mathcal{V}^{\Delta}(t,x) \leq G(t,\mathcal{V}(t,x)) \tag{3.6.3}$$

where $G \in C_{rd}[R^+ \times R^q_+, R^q]$, $G(t,w)$ is quasimonotone nondecreasing in w, and for each i, $1 \leq i \leq q$, $G_i(t,w)\mu^*(t) + w_i$ is nondecreasing in w for $t \in \mathbb{T}$.

(A5) *$\phi(\lambda) < A$ and $a(\lambda) < b(A)$ hold;*

then the practical stability properties of (3.6.1) *imply the corresponding (h_0,h)-practical stability properties of* (3.1.1).

Proof: We shall prove only (h_0,h)-asymptotic practical stability and (h_0,h)-strong practical stability of (3.1.1). For this purpose let us first prove

(h_0,h)-practical stability. Since the system (3.6.1) is practically stable with respect to $(a(\lambda), b(A))$ we have

$$\sum_{i=1}^{q} w_{0_i} < a(\lambda) \quad \text{implies} \quad \sum_{i=1}^{q} w_i(t, t_0, u_0) < b(A),\ t \geq t_0 \tag{3.6.4}$$

We claim that the system (3.1.1) is (h_0,h)-practically stable with respect to (λ,A). If this were not true there would exist a $t_1 > t_0$ and a solution $x(t) = x(t, t_0, x_0)$ of (3.1.1) such that

$$h_0(t_0,x_0) < \lambda\,,\quad h(t_1,x(t_1))= A,\ \text{and}\ h(t, x(t)) < A\ \textit{for}\ t_0 \leq t < t_1 \tag{3.6.5}$$

In view of conditions (A2) and (A5), we have $h(t_0,x_0) \leq \phi(h_0(t_0,x_0)) < \phi(\lambda) < A$. Then, setting $u_{0i} = V_i(t_0, x_0)$, we get because of Theorem 3.1.1 and Remark 3.1.1 the estimate

$$V(t, x(t)) \leq r(t, t_0, u_0), \tag{3.6.6}$$

where $r(t,t_0,u_0)$ is the maximal solution of (3.6.1) with $u_{0i} = V_i(t_0, x_0)$. Since

$$\sum_{i=1}^{q} w_{0_i} < \sum_{0}^{\infty} V_i(t_0,x_0) \leq a(h_0(t_0,x_0)) < a(\lambda),$$

it then follows from (3.6.4)-(3..7.6) that

$$b(A) = b(h(t_1,x(t_1)) \leq \sum_{i=1}^{q} V_i(t_1,x(t_1)) \leq \sum_{i=1}^{q} r_i(t_1, t_0, u_0) < b(A).$$

Since this is a contradiction, system (3.1.1) is (h_0,h)-practically stable.

Suppose next that (3.6.1) is practically quasi stable. Then, given $a(\lambda) > 0$, $\epsilon > 0$, $t_0 \in \mathbb{T}$, there exists $T = T(t_0,\epsilon,a(\lambda)) > 0$ such that

$$\sum_{i=1}^{q} w_{0_i} < a(\lambda) \quad \text{implies} \quad \sum_{i=1}^{q} w_i(t, t_0, u_0) < b(\epsilon), \qquad t \geq t_0 + T \tag{3.6.7}$$

Let $h_0(t_0,x_0) < \lambda$ and set $w_{0i} = V_i(t_0,x_0)$, then the estimate obtained in (3.6.6) is valid for $t \geq t_0$. Now suppose that there exists a sequence $\{t_k\}_{k\geq 1}$, $t_k \geq t_0 + \mathrm{T}$, $t_k \to \infty$, such that $b(\epsilon) \leq \sum_{i=1}^{q} w_i(t_k, t_0, u_0)$ where $w(t, t_0, u_0)$ is any solution of (3.6.1) such that $\sum_{i=1}^{q} w_{0_i} < a(\lambda)$. Then, in view of (3.6.6)-(3.6.7), we have a contradiction since

$$b(\epsilon) \leq \sum_{i=1}^{q} V_i(t_k,x(t_k)) \leq \sum_{i=1}^{q} r_i(t_k, t_0, u_0) < b(\epsilon).$$

Hence, (3.1.1) is (h_0,h)-practically asymptotically stable.

Next, we prove (h_0,h)-strong practical stability of (3.1.1). For that purpose assume that (3.6.1) is strongly practically stable for $(a(\lambda), b(A), b(B), \mathrm{T}) > 0$. Then, we need to prove only (h_0,h)-practical quasi-stability of (3.1.1). Since (3.6.1) is practically quasi-stable, we have for some $t_0 \in \mathbb{T}$

$$\sum_{i=1}^{q} w_{0_i} < a(\lambda) \quad \text{implies} \quad \sum_{i=1}^{q} w_i(t, t_0, u_0) < b(B), \qquad t \geq t_0 + T$$

where $w(t,t_0,w_0)$ is any solution of (3.6.1). Suppose that $h_0(t_0,x_0) < \lambda$ so that $h(t,x(t)) < A$ for $t \geq t_0$. This is possible because of the practical stability of (3.6.1). Then, with $w_{0i} = V_i(t_0, x_0)$ the estimate given in (3.6.6) applies. This implies

$$b(h(t,x(t)) \leq \sum_{i=1}^{q} V_i(t,x(t)) \leq \sum_{i=1}^{q} r_i(t, t_0, u_0) < b(B), \quad t \geq t_0 + T.$$

Hence, $h(t,x(t)) < B$ for $t \geq t_0 + T$ whenever $h_0(t_0, x_0) < \lambda$, and therefore (3.1.1) is (h_0,h)-strongly practically stable. Similarly one can prove other (h_0,h) uniform and nonuniform practical stability properties of (3.1.1). Hence, the proof of the theorem is complete.

An example of Theorem 3.6.2 for the case of (h_0,h)-practical stability which will be useful in the proof of Theorem 3.6.2 is the case where $G(t,u)$ takes the form $\epsilon(t)$ with $\epsilon \in C_{rd}[\mathbb{T}, R_+^q]$, $\epsilon_i \in L^1$, $1 \leq i \leq q$. Then the solution of (3.6.2) is

$$w(t) = w(t_0) + \int_{t_0}^{t} \epsilon(s)ds, \quad t \geq t_0$$

which, in view of the assumption on ϵ_i, yields, for $1 \leq i \leq q$,

$$w_i(t) \leq w_i(t_0) + \int_{t_0}^{\infty} \epsilon_i(s)ds = w_i(t_0) + \sigma_i(t_0), \quad t \geq t_0$$

where $\sigma_i(t_0) \equiv \int_{t_0}^{\infty} \epsilon_i(s)ds$. Given $0 < \lambda < A$, assume that $h_0(t_0, x_0) < \lambda$ and set $w_{i0} = V_i(t_0, x_0)$. Then,

$$\sum_{i=1}^{q} V_i(t,x(t)) \leq \sum_{i=1}^{q} V_i(t_0,x_0) + \sum_{i=1}^{q} \sigma_i(t_0)$$

which in view of (3.6.2) yields

$$\sum_{i=1}^{q} V_i(t,x(t)) \leq a(h_0(t_0, x_0)) + \sum_{i=1}^{q} \sigma_i(t_0) < a(\lambda) + \sum_{i=1}^{q} \sigma_i(t_0).$$

For the comparison system to be practically stable we have to show that there exists at least a $t_0 \in \mathbb{T}$ such that $a(\lambda) + \sum_{0}^{q} \sigma_i(t_0) < b(A)$. In view of assumption (A4) of Theorem 3.6.1 and the assumption on ϵ_i, it is clear that such a t_0 exists. Hence, by Theorem 3.6.1, system (3.1.1) is (h_0,h)-practically stable.

Next we prove a special case of (h_0,h)-practical asymptotic stability which cannot be proved using Theorem 3.6.1.

Theorem 3.6.2: *Assume that conditions (A1)-(A4) of Theorem* 3.6.1 *hold except for equation* (3.6.3) *which is modified as follows*

$$D^+V_j^{\Delta}(t,x) \leq -\gamma(h(t,x)) + \epsilon_j(t), \qquad 1 \leq j \leq q \tag{3.6.8}$$

$$D^+V_i^{\Delta}(t,x) \leq \epsilon_i(t), \qquad 1 \leq i \leq q, \ i \neq j$$

where $\gamma \in \mathcal{K}$, $\epsilon \in C_{rd}[\mathbb{T}, R_+^q]$, $\epsilon_i \in L^1$, $1 \leq i \leq q$. Assume further that $h(t,x)$ is locally Lipschitzian in x and that $D^+h^{\Delta}(t,x)$ is bounded on $\mathbb{T} \times R^n$, then system (3.1.1) *is (h_0,h)-practically asymptotically stable.*

Proof: By Theorem 3.6.1 it follows that system (3.1.1) is (h_0,h)-practically stable. It is now left to show that any solution $x(t)$ of (3.1.1) with $h_0(t_0,x) < \lambda$ satisfies $h(t,x(t)) \to 0$ as $t \to \infty$. Suppose that $h(t_0,x_0) < \lambda$. First, note that $\liminf_{t \to \infty} h(t,x(t)) = 0$, for otherwise in view of (3.6.8) above we would have $V(t,x(t)) \to -\infty$ *as* $t \to +\infty$. Next, suppose that $\limsup_{t \to \infty} h(t,x(t)) \neq 0$. Then, for any $\eta > 0$, there exists a divergent sequence $\{t_i\}$, $t_i \in \mathbb{T}$, $t_i \to \infty$ as $i \to \infty$ such that $h(t_i,x(t_i) \geq \eta$. Each t_i can be one of the following

(i) left-scattered, right-scattered (ls-rs),

(ii) left-dense, right dense (ld-rd),

(iii) left-dense, right-scattered (ld-rs),

(iv) left-scattered, right-dense (ls-rd).

Without loss of generality we can assume that there exists a subsequence $\{t_k\}$ of $\{t_i\}$, satisfying $t_k \to \infty$ as $k \to \infty$ and $h(t_k,x(t_k)) \geq \eta$, such that one of the following cases occurs.

Case 1: All t_k, $k = 1,2,\ldots$, are ls-rs. This case corresponds to the discrete topology. Inequality (3.6.8) takes the form

$$D^+V_j^{\Delta}(t) = \frac{V_j(\sigma(t_k), x(\sigma(t_k))) - V_j(t_k, x(t_k))}{\mu^*(t_k)} \leq -\gamma(h(t_k,x(t_k)) + \epsilon_{j,k}.$$

which yields

$$0 < \mathcal{V}_j(\sigma(t_k), x(\sigma(t_k))) \leq \mathcal{V}_j(t_k, x(t_k)) + (-\gamma(h(t_k, x(t_k)) + \epsilon_{j,k})\, \mu^*(t_k).$$

But since for any $k > 0$

$$\mathcal{V}_j(t_k, x(t_k)) = \mathcal{V}_j(\sigma(t_{k-1}), x(\sigma(t_{k-1}))$$

$$\leq \mathcal{V}_j(\sigma(t_{k-1}), x(\sigma(t_{k-1})) + (-\gamma(h(t_{k-1}, x(t_{k-1})) + \epsilon_{j,k-1})\mu^*(t_k).$$

$$\leq \mathcal{V}_j(\sigma(t_{k-1}), x(\sigma(t_{k-1})) + (-\gamma(\eta) + \epsilon_{j,k-1})\, \mu^*(t_k).$$

Therefore, we arrive at, with $\beta = \max_i \mu^*(t_i)$,

$$0 < \mathcal{V}_j(\sigma(t_k), x(\sigma(t_k))) \leq \mathcal{V}_j(t_0, x(t_0)) - (t_{k+1} - t_0)\, \gamma(\eta) + \beta \sum_{i=0}^{k} \epsilon_{j,i}\,.$$

Since $\epsilon_j \in L^1$ means, in the case of the discrete topology, that $\sum_{i=0}^{\infty} \epsilon_{j,i} < \infty$,

the right hand side will be negative for k is sufficiently large. This contradiction proves that $\limsup_{t \to \infty} h(t, x(t)) = 0$.

In the other cases, we can find another subsequence $\{t_k^*\}$ of $\{t_i\}$, satisfying $t_k^* \to \infty$ as $k \to \infty$, such that either

(i) $t_k^* < t_k < t_{k+1}^*$, $k = 1, 2, 3, \ldots,$

$$h(t_k^*, x(t_k^*) = \tfrac{\eta}{2}, \quad h(t_k, x(t_k)) = \eta,$$

$$\tfrac{\eta}{2} < h(t, x(t)) < \eta \quad \textit{for } t \in (t_k^*, t_k), \ k = 1, 2, \ldots.$$

or (ii) $t_k < t_k^* < t_{k+1}^*$, $k = 1, 2, 3, \ldots,$

$$h(t_k, x(t_k)) = \eta, \quad h(t^*_{k+1}, x(t^*_{k+1})) = \frac{\eta}{2},$$

$$\frac{\eta}{2} < h(t, x(t)) < \eta \quad for\ t \in (t_k, t^*_{k+1}), \ k = 1, 2, \ldots.$$

or both. Suppose that $D^+h(t,x(t)) \leq M$ then it is easy to obtain for case (i), using the inequality $t_k - t^*_k > \eta/2M$, that

$$V_j(t_n, x(t_n)) \leq V_j(t_0, x_0) - \sum_{k=1}^{n} \int_{t^*_k}^{t_k} \gamma(h(s, x(s))\, ds + \int_{t_0}^{\infty} \epsilon_j(s)\, ds$$

$$\leq V_j(t_0, x_0) + \sigma_j(t_0) - n\, \gamma(\tfrac{\eta}{2}) . \tfrac{\eta}{2M}$$

which can be made negative for n sufficiently large. Hence, we have a contradiction , and $\limsup_{t \to \infty} h(t,x(t)) = 0$. A similar result can be proven for case (ii) using $D^+h(t,x(t)) \geq -N$ and integrating from t_k to t^*_{k+1} in the second integral above. Therefore, $\lim_{t \to \infty} h(t,x(t)) = 0$ in all 3 cases and this completes the proof.

3.7 Notes

Comparison theorems described in terms of Lyapunov-like functions in section 3.1 are adapted from the work of Kaymakcalan [22] which are based on the corresponding results for continuous and discrete dynamic systems in [20, 25, 31, 34, 36, 40]. Stability theory developed in Sections 3.2 and 3.3 is based on the work in [22, 26]. the presentation of the material in these sections employs necessary refinements and corrections. Section 3.4

describes the stability results of conditionally invariant sets which can be applied to control uncertain dynamic systems. See for details Drici and Lakshmikantham [14] and Lakshmikantham and Sivasundaram [39]. See Kaymakcalan [24] for the results in Section 3.6 that deal with stability criteria relative to two measures. See Lakshmikantham and Liu [34] for the theory of stability in terms of two measures. The method of vector Lyapunov functions and practical stability form the contents of Section 3.6 which is taken from Drici [12]. For Lyapunov stability theory for continuous and discrete dynamic systems see [10, 21, 34, 36, 37, 40].

Chapter 4

4.0 Introduction

We shall introduce, in this chapter, several possible areas of investigation of dynamic systems on time scales that await further progress. In Section 4.1, we shall develop the methods of upper and lower solutions and monotone iterative technique for initial and periodic boundary value problems. Section 4.2 is devoted to the method of generalized quasilinearization which offers not only monotone sequences that converge to the unique solution but also shows that the convergence is quadratic.

Section 4.3 is dedicated to the theory of fixed points which can be obtained as a result of monotone flows. Here the decomposition of a given function with the required monotone properties together with the existence of a suitable initial state plays an important role. We describe, in Section 4.4, results relative to invariant manifolds, while in Section 4.5, we consider uncertain dynamic systems and study practical stability criteria in terms of vector Lyapunov functions and two measures. In Section 4.6, we initiate the discussion of linear boundary value problems, and in Section 4.7, Sturmian theory for second order linear equations is developed. Finally, Section 4.8 is dedicated to the study of convexity of solutions relative to the initial data an Section 4.9, to the extension of LaSalle's Invariance Principle.

4.1 Monotone Iterative Technique

As is well known [29], the method of upper and lower solutions, coupled with the monotone iterative technique, manifests itself as an effective and flexible mechanism that offer theoretical, as well as constructive existence results in a closed set for differential equations. The upper and lower solutions that generate the closed set serve as upper and lower bounds for solutions which can be improved by monotone iterative procedures. Moreover, iteration schemes are also useful for the investigation of qualitative properties of solutions.

In this section, we shall investigate the method of upper and lower solutions applied to dynamic systems on times scales. We further employ monotone iterative technique for initial value problems and periodic boundary value problems in order to obtain extremal solutions.

Let us begin by considering the scalar initial value problem

$$u^{\Delta} = f(t,u),\, u(0) = u_0, \tag{4.1.1}$$

where $f \in C_{rd}[\mathbb{T} \times R, R]$ and $\mathbb{T}$ is a time scale such that $t = 0$ is the minimal element and $t = \mathbb{T}$ is the maximal element which is not left scattered.

Definition 4.1.1: A function $v \in C_{rd}[\mathbb{T}, R]$ which is differentiable is said to be an upper solution of (4.1.1) if

$$v^{\Delta} \geq f(t,v),\, t \in \mathbb{T},\, v(0) \geq u_0$$

and a lower solution of (4.1.1) if the reversed inequalities hold.

If we know the existence of upper and lower solutions w, v such that $v \leq w$, we can prove the existence of a solution in the closed set

$$\Omega = [(t,u) \colon v(t) \leq u \leq w(t),\, t \in \mathbb{T}].$$

In order to prove such a result we need the following.

Proposition 4.1.1: *Consider the system* (4.1.1) *where* $f \in C_{rd}[\mathbb{T} \times R^n, R^n]$ *and* $f(t,x)$ *is bounded everywhere. Then there exists a solution* $x(t)$ *on* $\mathbb{T}$ *of* (4.1.1).

The proof follows by the usual procedure employing Schauder fixed point theorem and we omit it. See [29] for corresponding theorem in differential equations.

Theorem 4.1.2: *Let the differentiable maps* $v, w \in C_{rd}[\mathbb{T}, R]$ *be lower and upper solutions of* (4.1.1) *such that* $v(t) \leq w(t)$ *on* $\mathbb{T}$ *and* $f \in C_{rd}[\Omega, R]$ *with* f *being bounded on* Ω. *Then there exists a solution* $u(t)$ *of* (6.1.1) *such that* $v(t) \leq u(t) \leq w(t)$ *on* $\mathbb{T}$, *provided that* $v(0) \leq u_0 \leq w(0)$.

Proof: Let $P: \mathbb{T} \times R \to R$ be defined by $P(t,u) = max[v(t), min(u, w(t))]$. Then $f(t, P(t,u))$ defines a rd-continuous extension of f to $\mathbb{T} \times R$ which is also bounded since f is bounded on Ω. Therefore by Proposition 4.1.1, $u^\Delta = f(t, P(t,u))$, $u(0) = u_0$ has a solution on $\mathbb{T}$. For $\epsilon > 0$ consider $w_\epsilon(t) = w(t) + \epsilon(1+t)$ and $v_\epsilon(t) = v(t) - \epsilon(1+t)$. We have $v_\epsilon(0) < u_0 < w_\epsilon(0)$. We wish to show that

$v_\epsilon(t) < u(t) < w_\epsilon(t)$ on $\mathbb{T}$. For this, we employ the induction argument.

Let $A(t)\colon v_\epsilon(t) < u(t) < w_\epsilon(t)$, $t \in \mathbb{T}$. If t is right-scattered, suppose that for $\sigma(t) \in \mathbb{T}$, we have $v_\epsilon(t) < u(t) < w_\epsilon(t)$ on $[0, \sigma(t))$ and $u(\sigma(t)) = w_\epsilon(\sigma(t))$. Then $u(\sigma(t)) > w(\sigma(t))$ and so $P(\sigma(t), u(\sigma(t))) = w(\sigma(t))$. Also,

$$v(\sigma(t)) \leq P(\sigma(t), u(\sigma(t))) \leq w(\sigma(t)).$$

Hence, by w being an upper solution,

$$w^\Delta(\sigma(t)) \geq f(\sigma(t), P(\sigma(t), u(\sigma(t))) = u^\Delta(\sigma(t)).$$

Since $w_\epsilon^\Delta(\sigma(t)) > w^\Delta(\sigma(t))$, we have $w_\epsilon^\Delta(\sigma(t)) > u^\Delta(\sigma(t))$ which implies

$$\frac{w_\epsilon(\sigma(t)) - w_\epsilon(\sigma(t) - \mu^*(t))}{\mu^*(t)} > \frac{u(\sigma(t) - u(\sigma(t) - \mu^*(t))}{\mu^*(t)}.$$

Since $u(\sigma(t)) = w_\epsilon(\sigma(t))$, we get

$$w_\epsilon(\sigma(t) - \mu^*(t)) < u(\sigma(t) - \mu^*(t))$$

which contradicts the assumption $u(t) < w_\epsilon(t)$ for $t \in [0, \sigma(t))$. Consequently, $u(\sigma(t)) < w_\epsilon(\sigma(t))$ and $A(\sigma(t))$ holds.

If t is right-dense, choosing a t_1 in a neighborhood U to the right of t, which replaces the role of $\sigma(t)$ in the right-scattered case, we arrive at the conclusion that $v_\epsilon(t) < u(t) < w_\epsilon(t)$ on all of U, hence $A(s)$ is true for all $s \in U$, $s \geq t$.

For a left-dense t, if $A(s)$ is true for all $s \in [0, t)$ since u, v, w are all rd-continuous the limit from left exists and $\lim_{s \to t^-} v_\epsilon(s) < \lim_{s \to t^-} u(s) < \lim_{s \to t^-} w_\epsilon(s)$ implies $v_\epsilon(t) \leq u(t) \leq w_\epsilon(t)$. Hence we have $A(t)$ is true on $\mathbb{T}$. Letting $\epsilon \to 0^+$ we get

$$v(t) \leq u(t) \leq w(t) \text{ on } \mathbb{T}.$$

The proof is therefore complete.

We shall next prove a method which yields monotone sequences that converge to solutions of (4.1.1). Since each member of these sequences happens to be the solution of a certain linear dynamic equation which can explicitly be computed, the advantage and importance of the technique needs no special emphasis.

Theorem 4.1.3: *Let $f \in C_{rd}[\mathbb{T} \times R, R]$, v_0, w_0 be lower and upper solutions of* (4.1.1) *such that $v_0 \leq w_0$ on $\mathbb{T}$. Suppose further that*

$$f(t,u) - f(t,\overline{u}\,) \geq -M(u - \overline{u}\,) \textit{ for } v_0(t) \leq \overline{u} \leq u \leq w_0(t) \quad (4.1.2)$$

with $M \geq 0$ *for* t *right dense and* $M \leq 0$ *for* t *right-scattered. Then there exist monotone sequences* $\{v_n\}$, $\{w_n\}$ *such that* $v_n \to v$, $w_n \to w$ *as* $n \to \infty$ *locally uniformly and monotonically on* $\mathbb{T}$ *and that* v, w *are minimal and maximal solutions of* $(4.1.1)$ *respectively.*

Proof: For any $\eta \in C_{rd}[\mathbb{T}, R]$ such that $v_0 \leq \eta \leq w_0$, we consider the linear differential equation

$$u^\Delta = f(t,\eta) - M(u-\eta), u(0) = u_0. \qquad (4.1.3)$$

It is clear that for every η, there exists a unique solution u of (4.1.3) on $\mathbb{T}$. Define a mapping A by $A\eta = u$. This mapping will be used to define the sequences $\{v_n\}$, $\{w_n\}$. Let us prove that,

(a) $v_0 \leq Av_0$, $w_0 \geq Aw_0$;

(b) A is a monotone operator on the segment

$$[v_0, w_0] = [u \in C_{rd}[\mathbb{T}, R]: v_0 \leq u \leq w_0].$$

To prove (a), set $Av_0 = v_1$ where v_1 is the unique solution of (4.1.3) with $\eta = v_0$. Setting $p = v_1 - v_0$, in view of (4.1.3) and v_0 being a lower solution, we see that,

$$p^\Delta = v_1^\Delta - v_0^\Delta \geq f(t,v_0) - M(v_1 - v_0) - f(t,v_0) = -Mp$$

$$\text{and } p(0) \geq 0.$$

Consequently, we get from Theorem 2.2.2 that $p(t) \geq 0$ on $\mathbb{T}$, which implies that $v_0 \leq Av_0$. Similarly, we can prove that $w_0 \geq Aw_0$.

To prove (b), let $\eta_1, \eta_2 \in [v_0, w_0]$ such that $\eta_1 \leq \eta_2$. Suppose that $u_1 = A\eta_1$ and $u_2 = A\eta_2$. Set $p = u_2 - u_1$ so that in view of (4.1.2) and (4.1.3)

$$\begin{aligned} p^\Delta &= f(t, \eta_2) - M(u_2 - \eta_2) - f(t, \eta_1) + M(u_1 - \eta_1) \\ &\geq -M(\eta_2 - \eta_1) - M(u_2 - \eta_2) + M(u_1 - \eta_1) = -Mp \end{aligned}$$

and $p(0) = 0$. As before, this implies that $A\eta_1 \leq A\eta_2$ proving (b).

We can now define the sequences $v_n = Av_{n-1}$, $w_n = Aw_{n-1}$ and conclude from the previous arguments that

$$v_0 \leq v_1 \leq v_2 \leq \ldots \leq v_n \leq w_n \leq \ldots \leq w_2 \leq w_1 \leq w_0 \text{ on } \mathbb{T}.$$

It then follows by standard arguments that $\lim_{n \to \infty} v_n = v$ and $\lim_{n \to \infty} w_n = w$ locally uniformly and monotonically on $\mathbb{T}$. It is easy to show that v and w are solutions of (4.1.1) in view of the fact that v_n, w_n satisfy

$$\begin{aligned} v_n^\Delta &= f(t, v_{n-1}) - M(v_n - v_{n-1}), v_n(0) = u_0, \\ w_n^\Delta &= f(t, w_{n-1}) - M(w_n - w_{n-1}), w_n(0) = u_0. \end{aligned}$$

To prove that v, w are minimal and maximal solutions of (4.1.1), we have to show that if u is any solution of (4.1.1) such that $v_0 \leq u \leq w_0$ on $\mathbb{T}$, then $v_0 \leq v \leq u \leq w \leq w_0$ on $\mathbb{T}$. To do this, suppose that for some n, $v_n \leq u \leq w_n$ on $\mathbb{T}$ and set $p = u - v_{n+1}$ so that

$$p^\Delta = f(t,u) - f(t,v_n) + M(v_{n+1} - v_n)$$

$$\geq -M(u - v_n) + M(v_{n+1} - v_n) = -Mp.$$

Since $p(0) = 0$, this implies $v_{n+1} < u$ on $\mathbb{T}$. Similarly, $u \leq w_{n+1}$ on $\mathbb{T}$ and hence $v_{n+1} \leq u \leq w_{n+1}$ on $\mathbb{T}$. Since $v_0 \leq u \leq w_0$ on $\mathbb{T}$, this proves by induction that $v_n \leq u \leq w_n$ on $\mathbb{T}$ for all n. Taking the limit as $n \to \infty$, we conclude $v \leq u \leq w$ on $\mathbb{T}$ and the proof is complete.

We shall next consider the periodic boundary value problem

$$u^\Delta = f(t,u), u(0) = u(T). \tag{4.1.4}$$

We give here the monotone technique which offers a method to obtain extremal solutions for periodic boundary value problem (4.1.4).

Lemma 4.1.1: *Let $m \in C_{rd}[\mathbb{T}, R]$ be differentiable. Then*

$$m^\Delta(t) \leq -Mm(t), \text{ on } \mathbb{T}, \text{ where } M > 0 \text{ and } m(0) \leq m(T) \tag{4.1.5}$$

imply $m(t) \leq 0$ on $\mathbb{T}$.

Proof: If the conclusion is false, then there exists a $t_0 \in \mathbb{T}$ and an $\epsilon > 0$ such that

$$m(t_0) = \epsilon,\ m(t) \leq \epsilon \text{ on } \mathbb{T}.$$

If $t_0 \in (0, T]$, then $m^\Delta(t_0) \geq 0$. Then by (4.1.5) it follows that $0 \leq m^\Delta(t_0) \leq -Mm(t_0) = -M\epsilon < 0$, which is a contradiction.

If $t_0 = 0$, then $m(T) \geq m(0) = \epsilon$, hence

$$0 \leq m^\Delta(T) \leq -Mm(T) \leq -M\epsilon < 0$$

which is a contradiction.

Hence $m(t) \leq 0$ on $\mathbb{T}$.

In order to develop the monotone method, we need the following assumptions:

(A_0)Let $v_0, w_0 \in C_{rd}[\mathbb{T}, R]$ be differentiable with $v_0 \leq w_0$ on $\mathbb{T}$ and (4.1.2) holds.

(A_1)(i) $v_0^\Delta \leq f(t, v_0)$, $t \in \mathbb{T}$ and $v_0(0) \leq v_0(T)$;

(ii) $w_0^\Delta \geq f(t, w_0)$, $t \in \mathbb{T}$ and $w_0(0) \geq w_0(T)$;

Theorem 4.1.4: *Let (A_0) hold. Then the condition (A_1) implies that there exist monotone sequences $\{v_n\}$, $\{w_n\}$ such that $v_n \to v$, $w_n \to w$*

as $n \to \infty$ uniformly and monotonically on $\mathbb{T}$ and that v, w are minimal and maximal solutions of (4.1.4) respectively.

Proof: For any $\eta \in [v_0, w_0] = [\eta \in C_{rd}[\mathbb{T}, R]\colon v_0(t) \leq \eta(t) \leq w_0(t)]$, consider the linear periodic boundary value problem

$$u^{\Delta} = G(t,u) = f(t,\eta(t)) - M(u-\eta(t)), u(0) = u(T). \quad (4.1.6)$$

This linear problem can be explicitly solved and a solution of (4.1.6) is given by

$$u(t) = \Phi_M(t)[u(0) + \int_0^t [f(s,\eta(s)) + M\eta(s)]\,\Phi_M(s)\,\Delta s],$$

where $\Phi_M(t)$ is the principal solution of $u^{\Delta} = -Mu$, with $\Phi(0) = \mathrm{I}$, and

$$u(0) = u(T) = \frac{1}{\Phi_M(t)-1}\int_0^T [f(s,\eta(s)) + M\eta(s)]\,\Phi_M(\mathrm{s})\Delta s.$$

We shall show that the solution $u(t)$ of (4.1.6) is unique. If not, let $u_1(t)$, $u_2(t)$ be two solutions of (4.1.6). Set $p(t) = u_1(t) - u_2(t)$. Then, $p^{\Delta} \leq -Mp$ and $p(0) = p(T)$. Now using Lemma 4.1.1, it follows that $u_1(t) \leq u_2(t)$. By a similar argument we can conclude $u_2(t) \leq u_1(t)$.

We now define a mapping $A\eta = u$, where u is the unique solution of (4.1.6) and show that

(a) $v_0 \leq Av_0$, $w_0 \geq Aw_0$,

(b) A possesses a monotone property on the segment $[v_0, w_0]$.

In order to prove (a) set $Av_0 = v_1$, where v_1 is the unique solution of the PBVP (4.1.6). Set $p(t) = v_0(t) - v_1(t)$. If $(A_1)(i)$ holds, then in view of (4.1.2) $p^\Delta = v_0^\Delta - v_1^\Delta \leq -Mp$ on $\mathbb{T}$ and $p(0) \leq p(T)$.

Hence using Lemma 4.1.1 $p(t) \leq 0$ on $\mathbb{T}$. This proves $v_0 \leq Av_0$. Similarly, we can show that $w_0 \geq Aw_0$.

To prove (b) let $\eta,\mu \in [v_0, w_0]$ such that $\eta \leq \mu$. Suppose that $u_1 = A\eta$ and $u_2 = A\mu$. Setting $p(t) = u_1(t) - u_2(t)$, we see that, using (4.1.2)

$$\begin{aligned} p^\Delta = u_1^\Delta - u_2^\Delta &= f(t,\eta) - M(u_1 - \eta) - f(t,\mu) + M(u_2 - \mu) \\ &\leq M(u - \eta) - M(u_1 - u_2) + M(\eta - \mu) \\ &= -Mp \end{aligned}$$

and $p(0) = p(T)$. Consequently, Lemma 4.1.1 gives $u_1 \leq u_2$ on $\mathbb{T}$ proving (b).

It is now easy to define the sequences $\{v_n(t)\}$, $\{w_n(t)\}$ such that $v_n = Av_{n-1}$ and $w_n = Aw_{n-1}$ and we can conclude

$$v_0 \leq v_1 \leq v_2 \leq \ldots \leq v_n \leq w_n \leq \ldots \leq w_1 \leq w_0 \text{ on } \mathbb{T}.$$

It then follows using standard arguments that $\lim_{n \to \infty} v_n(t) = v(t)$ and $\lim_{n \to \infty} w_n(t) = w(t)$ uniformly and monotonically on $\mathbb{T}$. This implies that

$$v^{\Delta} = f(t, v) \quad v(0) = v(T)$$

$$w^{\Delta} = f(t, w) \quad w(0) = w(T).$$

We shall show that v, w are minimal and maximal periodic solutions respectively of (4.1.4).

Let us assume that for some integer $k > 0$, $v_{k-1} \leq u \leq w_{k-1}$ on $\mathbb{T}$, where u is any periodic solution of (4.1.4) such that $u \in [v_0, w_0]$. Then setting $p = v_k - u$, we get by use of (4.1.2)

$$p^{\Delta} = f(t, v_{k-1}) - M(v_k - v_{k-1}) - f(t, u)$$

$$\leq M(u - v_{k-1}) - M(v_k - v_{k-1}) = -Mp$$

and $p(0) = p(T)$. This implies by Lemma 4.1.1 $p(t) \leq 0$. Using similar arguments we get $u \leq w_k$ on $\mathbb{T}$. Since $v_0 \leq u \leq w_0$ on $\mathbb{T}$, it follows by induction that $v_n \leq u \leq w_n$ on $\mathbb{T}$ for all n. Hence we have

$v \leq u \leq w$ on $\mathbb{T}$, proving v, w are minimal and maximal periodic solutions of (4.1.4) respectively.

We shall next develop the monotone technic for systems of initial value problems. Usually, one needs to assume at least mixed monotone property to achieve this generalization. we shall show that even where we does not have any monotone property, we can construct monotone sequences to the systems by utilizing a method known as the method of mixed monotonicity.

We consider the initial value problem

$$x^{\Delta} = f(t, x),\ x(0) = x_0,\ t \in \mathbb{T}^k \tag{4.1.7}$$

where $f : \mathbb{T}^k \times R^n \to R^n$ and f is rd-continuous on $\mathbb{T}^k \times R^n$.

A map $x : \mathbb{T}^k \to R^n$ is a solution of IVP (4.1.7) if $x(t)$ is an antiderivative of $f(t, x(t))$ on $\mathbb{T}^k$ and satisfies $x(t_0) = x_0$.

Hereafter in this section we shall assume that

$\mathbb{T}^k = [0, T_0] \cap \mathbb{T}$, $T_0 > 0$

Theorem 4.1.5: *Assume that there exists a function $F(t, x, y)$ which is rd-continuous on $\mathbb{T}^k \times R^n \times R^n \to R^n$ such that*

(i) *$F(t, x, y)$ is nondecreasing in x and nonincreasing y for $t \in \mathbb{T}^k$,*

(ii) $F(t,x,x) = f(t,x)$,

(iii) $\alpha^{\Delta} \le F(t,\alpha,\beta), \beta^{\Delta} \ge F(t,\beta,\alpha)$, $\alpha(t) \le \beta(t)$,

$t \in \mathbb{T}^k$, *where* $\alpha, \beta : \mathbb{T}^k \to R^n$ *are differentiable for* $t \in \mathbb{T}^k$.

Then there exist monotone sequences $\{\alpha_n(t)\}, \{\beta_n(t)\}$ *such that* $\alpha_n \to \rho$, $\beta_n \to r$ *as* $n \to \infty$ *locally, uniformly and monotonically on* $\mathbb{T}^k$ *and that* ρ, r *satisfy*

$$\rho^{\Delta} = F(t,\rho,r), \ \rho(0) = x_0 \tag{4.1.8}$$

$$r^{\Delta} = F(t,r,\rho), \ \ r(0) = x_0 \tag{4.1.9}$$

If in addition we assume

$$|F(t,xy) - F(t,y,x)| \le L|x-y| \tag{4.1.10}$$

then $x(t) = \rho(t) = r(t)$ *is the unique solution of* (4.1.7) *on* $\mathbb{T}^k$.

Proof: We construct $\alpha_n(t), \beta_n(t)$ as follows

$$\alpha^{\Delta}_{n+1} = F(t,\alpha_n,\beta_n), \ \ \alpha_{n+1}(0) = x_0\,, \tag{4.1.11}$$

$$\beta^{\Delta}_{n+1} = F(t,\beta_n,\alpha_n)\,, \beta_{n+1}(0) = x_0\,, \tag{4.1.12}$$

where $\alpha_0 = \alpha$, $\beta_0 = \beta$.

Since $\alpha^{\Delta}_1 = F(t,\alpha_0,\beta_0) \ge \alpha^{\Delta}_0$ and by (iii), it is clear that

$\alpha_0(t) \le \alpha_1(t)$, $\beta_0(t) \ge \beta_1(t)$ on $\mathbb{T}^k$

in view of the fact $\alpha_0(0) \le x_0 \le \beta_0(0)$.

Assume that for some $k > 1$,

$$\alpha_{k-1}(t) \leq \alpha_k(t) \ , \ \beta_k(t) \leq \beta_{k-1}(t) \text{ on } \mathbb{T}^k.$$

Then using (i) we get

$$\alpha_{k+1}^{\Delta} = F(t, \alpha_k(t), \beta_k(t)) \geq F(t, \alpha_{k-1}(t), \beta_{k-1}(t)) = \alpha_k^{\Delta}(t),$$

$$\alpha_{k+1}(0) = \alpha_k(0) = x_0;$$

$$\beta_{k+1}^{\Delta} = F(t, \beta_k(t), \alpha_k(t)) \leq F(t, \beta_{k-1}(t), \alpha_{k-1}(t)) = \beta_k^{\Delta}(t),$$

$$\beta_{k+1}(0) = \beta_k(0) = x_0;$$

which implies

$$\alpha_k(t) \leq \alpha_{k+1}(t) \ , \ \beta_{k+1}(t) \leq \beta_k(t) \ , \ t \in \mathbb{T}^k.$$

Hence by induction we have

$$\alpha_0 \leq \alpha_1 \leq \alpha_2 \leq \ldots \leq \alpha_n \leq \beta_n \leq \ldots \leq \beta_1 \leq \beta_0, \text{ on } \mathbb{T}^k.$$

It then follows by the standard argument that

$$\lim_{n \to \infty} \alpha_n(t) = \rho(t), \quad \lim_{n \to \infty} \beta_n(t) = r(t)$$

locally, uniformly and monotonically on $\mathbb{T}^k$.

In view of (4.1.11), (4.1.12), it is easy to see that $\rho(t), r(t)$ satisfy (4.1.8), (4.1.9).

By setting $M = |r - \rho|$,

$$M^{\Delta} \leq |r^{\Delta} - \rho^{\Delta}| = |F(t, r, \rho) - F(t, \rho, r)| \leq |r - \rho| = LM$$

Since $M(0) = 0$, we get $M(t) \leq 0$ which implies $x = \rho = r$ on $\mathbb{T}^k$.

Since $F(t, x, x) = f(t, x)$ we see that $x = \rho = r$ is the unique solution of (4.1.7) on $\mathbb{T}^k$ which completes the proof of the theorem.

Let us give some sufficient conditions for the system (4.1.7) such that F exists satisfying the needed conditions of Theorem 4.1.5.

Suppose that f can be split as a difference of two nondecreasing functions, namely

$$f(t, x) = f_1(t, x) - f_2(t, x)$$

then F in Theorem 4.1.5 can be taken as

$$F(t, x, y) = f_1(t, x) - f_2(t, y).$$

clearly F satisfies (i) and (ii). To obtain (iii) we need to assume that

$$\alpha^{\Delta} \leq f_1(t, \alpha) - f_2(t, \beta),$$

$$\beta^{\Delta} \geq f_1(t, \beta) - f_2(t, \alpha),$$

If f satisfies a Lipschitz condition, then F verifies (4.1.10).

Next result offers another set of sufficient conditions.

Theorem 4.1.6: *Assume that*

(i) $\alpha^{\Delta} \leq f(t, \alpha) - A(\beta - \alpha),$

$$\beta^{\Delta} \geq f(t, \beta) + A(\beta - \alpha),\ \alpha(t) \leq \beta(t), t \in \mathbb{T}^k$$

where $\alpha, \beta : \mathbb{T}^k \to R^n$ *are differentiable for* $t \in \mathbb{T}^k$, *(ii)* $-A(x-y) \leq f(t,x) - f(t,y) \leq A(x-y)$ *whenever* $\alpha(t) \leq y \leq x \leq \beta(t)$. *A being an* $n \times n$ *matrix of nonnegative elements.*

Then the system (4.1.7) *admits the method of mixed monotony.*

Proof: We define

$$F(t,y,z) = \tfrac{1}{2}\left[\, f(t,y) + f(t,z) + A(y-z)\right] \tag{4.1.13}$$

It is easy to see that $F(t,y,z)$ is mixed monotone and note that

$$-A(z-\overline{z}\,) \leq F(t,y,z) - F(t,\overline{y}\,,\overline{z}\,) \leq A(y-\overline{y}\,) \tag{4.1.14}$$

whenever $\alpha(t) \leq \overline{y} \leq y \leq \beta(t)$ and $\alpha(t) \leq \overline{z} \leq z \leq \beta(t)$.

In particular, we have

$$F(t,y,z) - F(t,z,y) = A(y-z)$$

From (4.1.14) we see that,

$$-A(\beta-\alpha) \leq F(t,\alpha,\beta) - f(t,\alpha)$$

which yields

$$\alpha^{\Delta} \leq f(t,\alpha) - A(\beta-\alpha) \leq F(t,\alpha,\beta).$$

Similarly, we get

$$\beta^{\Delta} \geq F(t,\beta,\alpha).$$

Finally $F(t,x,x) = f(t,x)$ follows from (4.1.13) by setting $x = y = z$ and the proof is complete.

Remark: Let us assume f is autonomous, and consider both discrete and continuous cases which are included in Theorem 4.1.5.

Case 1: If $\mathbb{T}^k = Z$ and $T_0 = n$, where $n > 0$ is some positive integer, then original problem reduces to the following result.

Theorem 4.1.7: *Assume that there exist a function $F(x,y)$ which is continuous on $R^n \times R^n \to R^n$ such that*

(i) $F(x,y)$ *is nondecreasing in x and nonincreasing in y,*

(ii) $F(x,x) = f(x)$,

(iii) $0 \leq F(\alpha,\beta)$, $0 \geq F(\beta,\alpha)$, *where α, β are such that $\alpha \leq \beta$.*

Then there exists monotone sequences $\{\alpha_n\}, \{\beta_n\}$ such that $\alpha_n \to \rho, \beta_n \to r$ as $n \to \infty$ monotonically ρ, r satisfy

$$0 = F(\rho, r),$$

$$0 = F(r, \rho).$$

If in addition we assume that

$$|F(x,y) - F(y,x)| \leq |x - y|$$

then $x = \rho = r$ *is the unique solution of* $0 = f(x)$.

Case 2: If $\mathbb{T}^k = [0, T\ ]$, then the problem reduces to the result of [ ].

4.2 Method of Quasilinearization

The well known method of quasilinearization [7, 8] has recently been generalized and extended [30, 31, 35, 48] so as to apply to a large class of functions. In this section, we develop a method which isolates the ideas imbedded in the method of quasilinearization so that it can be applied to a variety of problems. We shall show that the generalized method of quasilinearization not only offers monotone sequences that converge to the unique solution but also guarantees that the convergence is rapid.

Let us begin by considering the initial value problem for dynamic system on time scale

$$x^{\Delta} = f(t, x),\ x(0) = x_0, \tag{4.2.1}$$

where $f \in C_{rd}[\Omega_0, R^n]$, $\Omega_0 = [(t, x) : \alpha_0(t) \leq x \leq \beta_0(t),\ t \in \mathbb{T}]$ and $\alpha_0, \beta_0 \in C_{rd}[\mathbb{T}, R^n]$ such that $\alpha_0(t) \leq \beta_0(t)$, $t \in \mathbb{T}$. As usual, the time scale $\mathbb{T}$ is supposed to have $t = 0$ as a minimal element and $t = T$ as a maximal element which is not left scattered. Let $\Omega = [\,(t,x,y):$ $\alpha_0(t) \le x, y \le \beta_0(t),\ t \in \mathbb{T}\,]$. We can prove the following result.

Theorem 4.2.1: *Assume that there exists a function $S \in C_{rd}[\,\Omega, R^n]$ such that*

(i) $\alpha_0, \beta_0 \in C_{rd}[\,\mathbb{T}, R^n]$ *with* $\alpha_0(t) \le \beta_0(t)$ *and*

$$\alpha_0^{\Delta} \le S(t,\alpha_0,\alpha_0)\ ,\ \beta_0^{\Delta} \ge S(t,\beta_0,\beta_0);$$

(ii) S *is bounded on* Ω, $S(t,x,y)$ *is quasimonotone nondecreasing in* x *and* $S_i(t,x,y)\mu^*(t) + x_i$ *is nondecreasing in* x_i *for each* $1 \le i \le n$, $t \in \mathbb{T}$ *and* $\alpha_0(t) \le y \le \beta_0(t)$;

(iii) $S(t,x,x) = f(t,x)$ *in* Ω, *and* $S(t,x,y) \le S(t,x,x)$, $x \ge y$, $t \in \mathbb{T}$;

(iv) $S_i(t,x_1,y) - S_i(t,x_2,y) \le L\sum_{i=1}^{n}(x_{1i} - x_{2i})$ *whenever* $x_1 \ge x_2$ *and* $\alpha_0(t) \le y \le \beta_0(t)$, $t \in \mathbb{T}$;

(v) $S(t,x,x) - S(t,y,z) \le N(x-y)^{1+\alpha} + M(x-z)$, *whenever* $\beta_0 \ge x \ge y \ge x \ge \alpha_0(t), t \in \mathbb{T}$, *where* N, M *are* $n \times n$ *nonnegative matrices.*

Then there exists monotone sequence $\{\alpha_n(t)\}$ *which converges locally uniformly to the unique solution of (4.2.1) on* $\mathbb{T}$ *and the convergence is superlinear.*

Proof: Consider the IVP for dynamic system

$$\alpha_1^\Delta \leq S(t, \alpha_1, \alpha_0), \alpha_1(0) = x_0 \tag{4.2.2}$$

Now $\alpha_0^\Delta \leq S(t, \alpha_0, \alpha_0)$, and by (iii), $\beta_0^\Delta \geq S(t, \beta_0, \beta_0) \geq S(t, \beta_0, \alpha_0)$.

Hence by Theorem 4.1.2 there exists a solution $\alpha_1(t)$ of (4.2.2) such that $\alpha_0 \leq \alpha_1 \leq \beta_0$ on $\mathbb{T}$. An application of Theorem 2.2.2 shows that α_1 is the unique solution of (4.2.2). We note that, because of (iii), $\alpha_1^\Delta \leq S(t, \alpha_1, \alpha_0) \leq S(t, \alpha_1, \alpha_1)$.

Consider the IVP

$$\alpha_2^\Delta \leq S(t, \alpha_2, \alpha_1),\ \alpha_2(0) = x_0 \tag{4.2.3}$$

We have $\alpha_1^\Delta \leq S(t, \alpha_1, \alpha_1)$, and as before by (iii), $\beta_0^\Delta \geq S(t, \beta_0, \beta_0) \geq S(t, \beta_0, \alpha_1)$. Hence by Theorem 4.1.2 there exists a solution $\alpha_2(t)$ of (4.2.3) such that $\alpha_1 \leq \alpha_2 \leq \beta_0$ on $\mathbb{T}$. Again Theorem 2.2.2 shows that α_2 is the unique solution of (4.2.3). This implies

$$\alpha_0 \leq \alpha_1 \leq \alpha_2 \leq \beta_0 \text{ on } \mathbb{T}.$$

Proceeding successively, we obtain

$$\alpha_0 \leq \alpha_1 \leq \ldots\ldots \leq \alpha_n \leq \beta_0 \text{ on } \mathbb{T},$$

where α_n is the unique solution of

$$\alpha_n^\Delta \leq S(t, \alpha_n, \alpha_{n-1}),\ \alpha_n(0) = x_0 \text{ on } \mathbb{T}.$$

Employing standard arguments, we get $\lim_{n \to \infty} \alpha_n(t) = x(t)$ locally uniformly and monotonically on $\mathbb{T}$. It is then easy to see that $x(t)$ is the unique solution of

$$x^\Delta = S(t, x, x) = f(t, x).$$

To show that α_n converges to x superlinearly, let

$$p_{n+1}(t) = x(t) - \alpha_{n+1}(t),$$

so that $p_{n+1}(0)$=0, and

$$p_{n+1}^\Delta(t) = f(t, x(t)) - S(t, \alpha_{n+1}(t), \alpha_n(t))$$

$$= S(t, x(t), x(t)) - S(t, \alpha_{n+1}(t), \alpha_n(t)).$$

Hence by (iv), we get, noting that $p_n(t) \geq 0$,

$$p_{n+1}^\Delta(t) \leq M p_{n+1}(t) + N\, p_n^{1+\alpha}(t),\ \ p_{n+1}(0)=0.$$

Thus, we obtain

$$p_{n+1}(t) \leq N \int_0^t \Phi_M(t, \sigma(s)) p_n^{1+\alpha}(s)\, \Delta s,$$

using the variation of parameter formula, where $\Phi_M(t, s)$ is the fundamental matrix solution of

$$x^\Delta = M\, x,\ x(0) = I \text{ (identity matrix)}.$$

We then get

$$\max_{\mathbb{T}} |p_{n+1}(t)| \leq K \max_{\mathbb{T}} |p_n(t)|^{1+\alpha}$$

where $K = \int_0^T N|\Phi_M(T,\sigma(s))\Delta s|$, and the proof is complete.

To demonstrate the existence of such a function $S(t,x,y)$, we now give sufficient conditions on $f(t,x)$ such that $S(t,x,y)$ satisfies the properties of Theorem 4.2.1.

Assume that

$(C0)$ $\alpha_0, \beta_0 \in C_{rd}[\mathbb{T}, R]$ with $\alpha_0(t) \leq \beta_0(t)$ and $\alpha_0^{\Delta} \leq f(t,\alpha_0)$, $\beta_0^{\Delta} \geq f(t,\beta_0)$ on $\mathbb{T}$;

$(C1)$ f is bounded on $Q = [(t,x) : \alpha_0(t) \leq x \leq \beta_0(t),\ t \in \mathbb{T}]$, $f(t,x)\,\mu^*(t) + x$ is non decreasing in x for each $t \in \mathbb{T}$ and $f_x(t,x)$, $f_{xx}(t,x)$exists, rd-continuous on Q;

$(C2)$ $f_{xx}(t,x) \geq 0$ on Q.

Suppose that $(C0)$ and $(C1)$ hold, then in view of $f_{xx}(t,x) \geq 0$, we get for $x \geq y$

$$f(t,x) \geq f(t,y) + f_x(t,y)(x-y)$$

and hence if we set

$$S(t,x,y) \equiv f(t,y) + f_x(t,y)(x-y),$$

then it is easy to verify that $S(t,x,y)$ satisfies the conditions (i) to (iv) of Theorem 4.2.1 with $\alpha = 1$.

If on the other hand, $(C2)$ is weakened to

$(C2)^* \; f_{xx}(t,x) + \phi_{xx}(t,x) \geq 0$ where $\phi \in C_{rd}[Q, R]$, ϕ_x, ϕ_{xx} exists and rd-continuous and $\phi_{xx}(t,x) \geq 0$ on Q;

then we define

$$S(t,x,y) \equiv f(t,y) + [f_x(t,y) + \phi_x(t,y)](x-y) - \phi(t,x) + \phi(t.y),$$

it is not difficult to verify that this $S(t,x,y)$ also satisfies all the conditions of Theorem 4.2.1.

In order to improve Theorem 4.2.1 so as to obtain lower and upper monotone sequences which converge to the unique solution and the convergence is super linear, we need the following two results concerning the upper and lower solutions. Theorem 4.2.2 is a restatement of Theorem 2.2.2, and Theorem 4.2.3 is an extension of Theorem 4.1.2 for systems, whose proof is omitted since it is similar.

Theorem 4.2.2: *Let $v, w \in C^1_{rd}[\mathbb{T}, R^n]$ satisfying*

$$v^\Delta(t) \leq f(t, v(t)),\; w^\Delta(t) \geq f(t, w(t)), t \in \mathbb{T},$$

where $f \in C_{rd}[\mathbb{T} \times R^n, R^n]$, and $f(t,x)$ is quasimonotone nondecreasing in x, and for each i, $1 \leq i \leq n$, $f_i(t,x)\mu^*(t) + x_i$ is nondecreasing in x_i, for each $t \in \mathbb{T}$. Then $v(t_0) \leq w(t_0)$ implies that $v(t) \leq w(t), t \in \mathbb{T}$, provided f satisfies

$$f_i(t,x) - f_i(t,y) \le L\sum_{i=1}^{n}(x_i - y_i)\,,\ x \ge y.$$

Theorem 4.2.3: *Let $v, w \in C^1_{rd}[\mathbb{T}, R^n]$ be lower and upper solutions of* (4.2.1) *such that $v(t) \le w(t)$, on $\mathbb{T}$ and $f \in C_{rd}[\,\Omega, R^n]$ where $\Omega = [\,(t,u) : v(t) \le u(t) \le w(t),\, t \in \mathbb{T}]$, f is bounded on Ω, f is quasimonotone nondecreasing in x, $f_i(t,x)\mu^*(t) + x_i$ is nondecreasing in x_i, for each $t \in \mathbb{T}$. Then there exists a solution $u(t)$ of* (4.2.1) *such that $v(t) \le u(t) \le w(t)$, on $\mathbb{T}$, provided that $v(0) \le u_0 \le w(0)$.*

Let $\Omega = [\,(t,x,y,z) : \alpha_0(t) \le x, y, z \le \beta_0(t),\, t \in \mathbb{T}\,]$. Relative to (4.2.1), we now prove the following result.

Theorem 4.2.4: *Suppose that there exists a function $\alpha_0, \beta_0, S_j, j = 1,2$ such that the following assumptions are satisfied:*

$(H0)$ $S_j \in C_{rd}[\,\mathbb{T}, R^n]\,, \Omega = [\,(t,x,y,z) : t \in \mathbb{T}, \alpha_0(t) \le x, y, z \le \beta_0(t)\,], S_j$ bounded on Ω, $S_j(t,x,y,z)$ is quasimonotone non-decreasing in x and $S_j(t,x,y,z)\mu^*(t) + x_i$ is nondecreasing in x_i for each $1 \le i \le n$, and $\alpha_0(t) \le y, z \le \beta_0(t)$;

$(H1)$ $\alpha_0, \beta_0 \in C^1_{rd}[\,\mathbb{T}, R^n]\,, \alpha_0^\Delta \le S(t, \alpha_0, \alpha_0, \beta_0)\,,$ $\beta_0^\Delta \ge S(t, \beta_0, \alpha_0, \beta_0)\,$; and $\alpha_0(t) \le \beta_0(t)$ on $\mathbb{T}$.

$(H2)$ $\quad S_1(t,y,y,z) = f(t,y)\,, S_2(t,z,y,z) = f(t,z)$ *and* $S_j(t,x,x,x) = f(t,x)$;

$(H3)$ $\quad S_1(t,x,y,z) \leq S_1(t,x,x,z)$, *if* $y \leq x$, *for each* $\alpha_0(t) \leq z \leq \beta_0(t)$ *and* $S_2(t,x,y,z) \geq S_1(t,x,y,x)$, *if* $x \leq z$, *for each* $\alpha_0(t) \leq y \leq \beta_0(t)$;

$(H4)$ $\quad S_j(t,x_1,y,z) - S_j(t,x_2,y,z) \leq L\sum_{i=1}^{n}(x_{1i} - x_{2i})$

whenever $\alpha_0(t) \leq x_2 \leq x_1 \leq \beta_0(t)$, *for some* $L \geq 0$, *and for* $t \in \mathbb{T}$, $x,u,v,w \in \Omega |S_j(t,x,x,x) - S_j(t,u,v,w)| \leq M|x-u| + N[|x-v|^{1+\eta} + |x-w|^{1+\eta}]$, $0 < \eta \leq 1$, N, M *being nonnegative constant matrices.*

Then there exist monotone sequence $\{\alpha_n(t)\}, \{\beta_n(t)\}$ *which converge locally, uniformly to the unique solution of* (4.2.1) *on* $\mathbb{T}$ *and the convergence is superlinear.*

Proof: We consider the initial value problems

$$\alpha_1^{\Delta} = S_1(t,\alpha_1,\alpha_0,\beta_0),\ \alpha_1(0) = x_0 \tag{4.2.4}$$

$$\beta_1^{\Delta} = S_1(t,\beta_1,\alpha_0,\beta_0),\ \beta_1(0) = x_0 \tag{4.2.5}$$

where $\alpha_0(0) \leq x_0 \leq \beta_0(0)$. Because of $(H0)$ and $(H2)$, we have

$$\alpha_0^{\Delta} \leq f(t,\alpha_0) = S_1(t,\alpha_0,\alpha_0,\beta_0),$$

$$\beta_0^\Delta \geq f(t,\beta_0) = S_1(t,\beta_0,\alpha_0,\beta_0).$$

Using Theorems 4.2.2 and 4.2.3, we get the existence of a unique solution $\alpha_1(t)$ of (4.2.4) on $\mathbb{T}$ satisfying $\alpha_0(t) \leq \alpha_1(t) \leq \beta_0(t)$ on $\mathbb{T}$.

Similarly, in view of $(H0)$ and $(H2)$, we also have

$$\alpha_0^\Delta \leq f(t,\alpha_0) = S_2(t,\alpha_0,\alpha_0,\beta_0),$$

$$\beta_0^\Delta \geq f(t,\beta_0) = S_2(t,\beta_0,\alpha_0,\beta_0),$$

which by Theorems 4.2.2 and 4.2.3, imply the existence of a unique solution $\beta_1(t)$ of (4.2.5) such that $\alpha_0(t) \leq \beta_1(t) \leq \beta_0(t)$ on $\mathbb{T}$.

Now, since $\alpha_0 \leq \alpha_1$ and $\beta_1 \leq \beta_0$, we see using $(H3)$,

$$\alpha_1^\Delta = S_1(t,\alpha_1,\alpha_0,\beta_0) \leq S_1(t,\alpha_1,\alpha_1,\beta_0),$$

$$\beta_1^\Delta = S_1(t,\beta_1,\alpha_0,\beta_0) \leq S_2(t,\beta_1,\alpha_0,\beta_1),$$

Hence by Theorem 4.2.2 , $\alpha_1(t) \leq \beta_1(t)$ on $\mathbb{T}$ and thus,

$$\alpha_0 \leq \alpha_1 \leq \beta_1 \leq \beta_0 \text{ on } \mathbb{T}. \tag{4.2.6}$$

Assume now that for some $k > 1$ and $t \in \mathbb{T}$,

$$\alpha_k^\Delta \leq f(t,\alpha_k)\,,\ \beta_k^\Delta \geq f(t,\beta_k),\ \alpha_k(t) \leq \beta_k(t)\,,$$

and consider the initial value problems

$$\alpha_{k+1}^\Delta = S_1(t,\alpha_{k+1},\alpha_k,\beta_k),\ \alpha_{k+1}(0) = x_0 \tag{4.2.7}$$

$$\beta_{k+1}^{\Delta} = S_2(t, \beta_{k+1}, \alpha_k, \beta_k),\ \beta_{k+1}(0) = x_0 \tag{4.2.8}$$

Clearly, $\alpha_k^{\Delta} \leq f(t, \alpha_k) = S_1(t, \alpha_k, \alpha_k, \beta_k)$, and $\beta_k^{\Delta} \geq f(t, \beta_k) = S_1(t, \beta_k, \alpha_k, \beta_k)$.

One obtains therefore by Theorems 4.2.2 and 4.2.3, the existence of a unique solution $\alpha_{k+1}(t)$ of (4.2.7) satisfying

$$\alpha_k(t) \leq \alpha_{k+1}(t) \leq \beta_k(t) \text{ on } \mathbb{T}.$$

Similar reasoning shows that there exists a unique solution $\beta_{k+1}(t)$ of (4.2.8) such that $\alpha_k(t) \leq \beta_{k+1}(t) \leq \beta_k(t)$, $t \in \mathbb{T}$. Using $(H3)$, together with the facts $\alpha_k \leq \alpha_{k+1}$, $\beta_{k+1} \leq \beta_k$, we see that

$$\alpha_{k+1}^{\Delta} = S_1(t, \alpha_{k+1}, \alpha_k, \beta_k) \leq S_1(t, \alpha_{k+1}, \alpha_{k+1}, \beta_k)$$

$$\beta_{k+1}^{\Delta} = S_2(t, \beta_{k+1}, \alpha_k, \beta_k) \geq S_2(t, \beta_{k+1}, \alpha_k, \beta_{k+1})$$

Hence by Theorem 4.2.2 yields $\alpha_{k+1}(t) \leq \beta_{k+1}(t)$ on $\mathbb{T}$. We have for $t \in \mathbb{T}$,

$$\alpha_k \leq \alpha_{k+1} \leq \beta_{k+1} \leq \beta_k. \tag{4.2.9}$$

By induction, we then have for all n,

$$\alpha_0 \leq \alpha_1 \leq \alpha_2 \leq \ldots \leq \alpha_n \leq \beta_n \leq \ldots \leq \beta_1 \leq \beta_0,\ t \in \mathbb{T},$$

with $\alpha_{n+1}^{\Delta} = S_1(t, \alpha_{n+1}, \alpha_n, \beta_n)$, $\beta_{n+1}^{\Delta} = S_2(t, \beta_{n+1}, \alpha_n, \beta_n)$, $\alpha_{n+1}(0) = x_0$, $\beta_{n+1}(0) = x_0$.

Employing standard arguments, it is now easy to conclude that

$$\lim_{n\to\infty}\alpha_n(t)=\rho(t),\quad \lim_{n\to\infty}\beta_n(t)=r(t)$$

locally, uniformly and monotonically on $\mathbb{T}$ and ρ , r are solutions of (4.2.3). An application of Theorem 4.2.2 implies that $x(t)=\rho(t)=r(t)$ is the unique solution of (4.2.3) on $\mathbb{T}$.

To prove superlinear convergence of α_n, β_n to x , we set $p_{n+1}(t)=x(t)-\alpha_{n+1}(t)$, $q_{n+1}(t)=\beta_{n+1}(t)-x(t)$, so that $p_{n+1}(t)\geq 0$, $q_{n+1}(t)\geq 0$, $p_{n+1}(0)=0=q_{n+1}(0)$.

Using $(H4)$, we then obtain,

$$p^{\Delta}_{n+1}(t)\leq Mp_{n+1}(t)+N[\ |p_n(t)|^{1+\eta}+|q_n(t)|^{1+\eta}],t\in\mathbb{T}.$$

Using the variation of parameters formula, we now get

$$p_{n+1}(t)\leq N\textstyle\int_0^t\Phi_M(t,\sigma(s))\,[\ |p_n(s)|^{1+\eta}+|q_n(s)|^{1+\eta}]\Delta s,$$

where $\Phi_M(t,s)$ is the fundamental matrix solution of the matrix differential equation

$$X^{\Delta}=M\,X,\ X(0)=I\ \text{(identity matrix)}.$$

It then follows that

$$\max_{\mathbb{T}}\ |p_{n+1}(t)|\leq K\left[\max_{\mathbb{T}}\ (|p_n(t)|^{1+\eta}+|q_n(t)|^{1+\eta})\right]$$

Similarly one can obtain the estimate

$$\max_{\mathbb{T}}\ |q_{n+1}(t)|\leq K\left[\max_{\mathbb{T}}\ (|p_n(t)|^{1+\eta}+|q_n(t)|^{1+\eta})\right]$$

The proof is complete.

As an application of Theorem 4.2.4, we shall give below conditions that guarantee the existence of S_j satisfying the assumptions of Theorem 4.2.4.

Assume that

$(A0)$ $\alpha_0, \beta_0 \in C_{rd}[\mathbb{T}, R^n]$ with $\alpha_0^{\Delta} \leq f(t, \alpha_0)$, $\beta_0^{\Delta} \geq f(t, \beta_0)$ and $\alpha_0 \leq \beta_0$ on $\mathbb{T}$.

$(A1)$ $f \in C_{rd}[\mathbb{T} \times R^n, R^n]$, and on $\Omega_0 = [(t, x) : t \in \mathbb{T}, \alpha_0(t) \leq x \leq \beta_0(t)]$, $f(t,x)$ is quasimonotone nondecreasing in x. Further $f(t,x)$ is uniformly convex in x and $f_x(t,x)$ exists and

$$|f_x(t,x) - f_x(t,y)| \leq L|x-y|^{\alpha} \text{ for } 0 < \alpha \leq 1$$

where L is an $n \times n$ matrix and $|x| = (|x_1|, \ldots, |x_n|)$.

We now set

$$S_1(t,x,y,z) = f(t,x) + f_x(t,y)(x-y),$$

$$S_2(t,x,y,z) = f(t,z) + f_x(t,y)(x-z),$$

for (t,x,y,z) on Ω. We note that if $f(t,x)$ is not convex in x on Ω_0, then we can still construct S_1 and S_2 provided there exists a function $\phi(t,x)$ such that $f(t,x) + \phi(t,x)$ is convex on Ω_0. However we need an extra assumption on $f(t,x) + \phi(t,x)$ so that we constructed systems

S_1 , S_2 to be quasimonotone nondecreasing on Ω_0.In view of the following Lemma, it is easy to verify that the assumptions $(H0)$ to $(H4)$ of Theorem 4.2.4 are satisfied by $S_1(t,x,y,z)$ and $S_2(t,x,y,z)$.

Lemma 4.2.1: *Let $f \in C_{rd}[\mathbb{T} \times R^n, R^n]$ and $f_x(t,x)$ exists and continuous. If $f(t,x)$ is uniformly convex in x, then*

$$f(t+h) - f(x) \leq f_x(t,x)\,h,$$

where the inequalities are componentwise.

Proof: Since $f_x(t,x)$ exists and continuous, we have f

$$f(t,x+h) = f(t,x) + f_x(t,x)\,h + |h|\,\eta(t,x,h), \quad h, x \in R^n \qquad (4.2.10)$$

where $|\eta(t,x,h)| \to 0$ as $|h| \to 0$. Since $f(t,x)$ is convex,

$$f(t, \lambda x + (1-\lambda)z) \leq \lambda f(t,x) + (1-\lambda) f(t,z), \quad 0 < \lambda < 1. \qquad (4.2.11)$$

Set $z = x + \frac{h}{1-\lambda}$, $0 < \lambda < 1$, so that $x + h = \lambda x + (1-\lambda)z$. Consequently, in view of (4.2.10) and (4.2.11) we get

$$\lambda f(t,x) + (1-\lambda) f(t,z) \geq f(t, \lambda x + (1-\lambda)z)$$

$$= \lambda f(t,x) + (1-\lambda) f(t,x) + f_x(t,x)\,h + |h|\,\eta(t,x,h),$$

which implies

$$|h|\,\eta(t,x,h) \le (1-\lambda)\left[f(t,z) - f(t,x) - f_x(t,x)\,\tfrac{h}{1-\lambda}\right]$$

$$= (1-\lambda)\,\eta(t,x,\tfrac{h}{1-\lambda})\left|\tfrac{h}{1-\lambda}\right|$$

$$= \eta(t,x,\tfrac{h}{1-\lambda})|h|.$$

Thus, we have $\eta(t,x,h) \le \eta(t,x,\frac{h}{1-\lambda})$ for $\lambda \in (0,1)$. Now taking $\eta = \frac{y}{n}$, $y \in R^n$ and $\lambda = 1 - \frac{1}{n}$, we obtain

$$\eta(t,x,\tfrac{y}{n}) \le \eta(t,x,y),$$

which yields $\eta(t,x,y) \ge 0$, because $\lim_{n \to \infty} \eta(t,x,\frac{y}{n}) \to 0$.

The proof is complete in view of (4.2.10).

4.3 Monotone Flows and Stationary Points

Let $\mathbb{T}$ be a time scale with $t = 0$ as the minimal element and no maximal element.

Consider the IVP for autonomous dynamic system on time scale $\mathbb{T}$,

$$u^{\Delta}(t) = g(u(t)). \tag{4.3.1}$$

Having Theorem 2.2.2. at out disposal, it is now easy to prove the following result.

Theorem 4.3.1: *Assume that*

(a) $g \in C[R^d, R^d]$, $g(u)$ *is quasimonotone nondecreasing in* u, $g_i(u)u^*(t) + u_i$ *is nondecreasing in* u_i *for each* i *and for* $L > 0$,

$$g_i(u) - g_i(v) \leq L\sum_{i=1}^{d}(u_i - v_i) \text{ whenever } u \geq v;$$

(b) $0 \leq g(\alpha_0)$, $g(\beta_0) \leq 0$ *and* $\alpha_0 \leq \beta_0$.

Then

(i) $\alpha_0 \leq u(t,\alpha_0), u(t,\beta_0) \leq \beta_0, t \in \mathbb{T}$;

(ii) $u(t,u_0) \leq u(t,v_0)$ *for all* $t \in \mathbb{T}$ *whenever* $\alpha_0 \leq u_0 \leq v_0 \leq \beta_0$;

(iii) $u(t,\alpha_0)$ *is nondecreasing and* $u(t,\beta_0)$ *is nonincreasing in* $t \in \mathbb{T}$;

(iv) $\underline{u} = \lim_{t \to \infty} u(t,\alpha_0), \overline{u} = \lim_{t \to \infty} u(t,\beta_0)$ *exist and satisfy* $g(\overline{u}) = 0$, $g(\underline{u}) = 0$.

Proof: Since $0 \leq g(\alpha_0)$ and $u^\Delta(t,\alpha_0) = g(u(t,\alpha_0))$, it follows by Theorem 2.2.2 that $\alpha_0 \leq u(t,\alpha_0)$, $t \in \mathbb{T}$. Similarly by repeated application of Theorem 2.2.2, we get

$u(t,\beta_0) \leq \beta_0$ and $u(t,u_0) \leq u(t,v_0)$, $t \in \mathbb{T}$ whenever $\alpha_0 \leq u_0 \leq v_0 \leq \beta_0$, proving (i) and (ii). Since (4.3.1) is autonomous, the solutions of (4.3.1) are unique.

Set $T(t)\,u_0 = u(t, u_0)$ and let $t, s \in \mathbb{T}$. Note that $\sigma(t+s) = t+s$, if $t+s \in \mathbb{T}$, and $\sigma(t+s) = \tau > t+s$, $\tau \in \mathbb{T}$, if $t+s \notin \mathbb{T}$. We observe that, setting

$v(t, v_0) =$

$$u(\sigma(t+s), u_0),\;\; v(0, v_0) = u(\sigma(s), u_0) = T(\sigma(s))\, u_0.$$

Consequently,

$$T(\sigma(t+s))\, u_0 = u(\sigma(t+s), u_0)$$

$$= T(t)\, u(\sigma(s), u_0) = T(t)\, T(\sigma(s))\, u_0.$$

Thus the solutions $u(t, u_0)$ satisfy the semigroup property in any case. Hence, it is clear that (iii) is satisfies. As a consequence (iv) follows and the proof is complete.

Remarks: (1) If in Theorem 4.3.1, $g(u) = S(u) - u$, then $\underline{u}$, $\overline{u}$ are the fixed points of S.

(2) In the literature, difference equations are sometimes written in the form

$$u(t+1) = g(u(t)) \text{ or } u_{n+1} = g(u_n).$$

As one can notice, in the present framework we use the difference operator to represent difference equations. Consequently care must be taken when deducing the results corresponding to difference equations.

Consider next the dynamic system on time scale $\mathbb{T}$,

$$x^{\Delta} = H(x) \tag{4.3.2}$$

where $H \in C[R^n, R^n]$. Utilizing Theorem 4.3.1, we can deduce the following result.

Theorem 4.3.2: *Assume that*

(i) *there exists a function* $F \in C[R^n \times R^n, R^n]$ *such that* $F(x, x) = H(x)$*;*

(ii) $F(x, y)$ *is quasimonotone nondecreasing in* x *and monotone nonincreasing in* y*;*

(iii) *for each* i, $1 \leq i \leq n$, $F_i(x, y)\mu^*(t) + x_i$ *is monotone nondecreasing in* x_i *and for* $L > 0$,

$$F_i(x, y) - F_i(\overline{x}\ , y) \leq L\sum_{i=1}^{n}(x_i - \overline{x}\ _i) \text{ whenever } x \geq \overline{x}\ ;$$

(iv) $0 \leq F(x_0, y_0), F(y_0, x_0) \leq 0$ *and* $x_0 \leq y_0$.

Then

(a) $x_0 \leq x(t, x_0) \leq x(t, y_0) \leq \beta_0, t \in \mathbb{T}$*;*

(b) $x(t, x_0)$ *is nondecreasing in t,* $x(t, y_0)$ *is nonincreasing in t,*

$$\lim_{t \to \infty} x(t, x_0) = \overline{x}\ ,\ \lim_{t \to \infty} x(t, y_0) = \overline{y}\ \text{ exist, and}$$

(c) *if* $z(t, z_0)$ *is the solution of* (4.3.2) *with* $x_0 \le z_0 \le y_0$, *then* $x(t, x_0) \le z(t, z_0) \le x(t, y_0)$, $t \in \mathbb{T}$ *and if* $\overline{z} = \lim_{t \to \infty} z(t, z_0)$ *exists then* $\overline{x} \le \overline{z} \le \overline{y}$.

Proof: Define the function $g \in C[R^d, R^d]$, $d = 2n$, by

$$g\begin{pmatrix} x \\ y \end{pmatrix} = \begin{pmatrix} F(x, y) \\ F(y, x) \end{pmatrix}$$

and introduce the ordering in $R^d_+ = R^n_+ \times \{-R^n_+\}$ by

$$\begin{pmatrix} x \\ y \end{pmatrix} \le \begin{pmatrix} x_1 \\ y_1 \end{pmatrix} \text{ iff } x \le x_1 \text{ and } y \ge y_1, \tag{4.3.3}$$

or in other words, if $u = \begin{pmatrix} x \\ y \end{pmatrix}$, $v = \begin{pmatrix} x_1 \\ y_1 \end{pmatrix}$, $u, v \in R^d_+$, we write $u \le v$ iff (4.3.3) holds.

Define the reflection operator $P: R^d \to R^d$ by

$$P\begin{pmatrix} x \\ y \end{pmatrix} = \begin{pmatrix} y \\ x \end{pmatrix}.$$

We note that g satisfies the assumptions of Theorem 4.3.1 when we let $\alpha_0 = \begin{pmatrix} x_0 \\ y_0 \end{pmatrix}$, $\beta_0 = P\alpha_0$, $u = \begin{pmatrix} x \\ y \end{pmatrix}$, $T(t)\alpha_0 = u(t, \alpha_0) = \begin{pmatrix} x(t, x_0) \\ x(t, y_0) \end{pmatrix}$.

Consequently, the conclusions (a) and (b) follow by Theorem 4.3.1 Since $\beta_0 = P\alpha_0$, $u(t, \beta_0) = Pu(t, \alpha_0)$, we have $P\underline{u} = \overline{u}$ where $\underline{u} = \begin{pmatrix} \overline{x} \\ \overline{y} \end{pmatrix}$. To prove (c), we let $\begin{pmatrix} z_0 \\ z_0 \end{pmatrix} = u_0$ so that $\alpha_0 \leq u_0 \leq \beta_0$, which implies from (ii) of Theorem 4.3.1 that $u(t, \alpha_0) \leq u(t, u_0) \leq u(t, \beta_0)$, $t \in \mathbb{T}$ proving (c). The proof is therefore complete.

Remark: Let $F \in C^1[R^n \times R^n, R^n]$, all the elements of the matrix $F_y(x, y) \leq 0$ and all the off-diagonal elements of the matrix $F_x(x, y) \geq 0$, which are standard assumptions considered. Then F satisfies the conditions of Theorem 4.3.2.

We shall next extend the results obtained above to dynamic systems on time scales relative to a Banach space. For this purpose, we shall prove first a basic result relative to the system of dynamic inequalities in arbitrary cones.

Let K be a cone in E, a real Banach space, that is closed, convex subset of E such that $K \cap \{-K\} = 0$, where 0 denotes the null element of E. The cone K induces the order relation in E defined by

$$x \leq y,\ x, y \in E \text{ if and only if } y - x \in K.$$

Theorem 4.3.3: *Assume that*

(i) $\mathbb{T}$ *is a time scale as before and the cone* K *is a distance set;*

(ii) $v, w \in C_{rd}[\mathbb{T}, E]$ *are differentiable for each* $t \in \mathbb{T}$ *and satisfy*

$$v^{\Delta}(t) \leq f(t, v(t)),\ w^{\Delta}(t) \geq f(t, w(t)),\ t \in \mathbb{T}, \tag{4.3.4}$$

where $f \in C_{rd}[\mathbb{T} \times E, E]$, $f(t,x)$ *is quasimonotone nondecreasing in* x, *and* $f(t,x)\mu^*(t) + x$ *is non decreasing in* x *for each* $t \in \mathbb{T}$, *relative to* K;

(iii) $|f(t,x) - f(t,y)| \leq g(t, |x-y|)$, $x \in E - K$ *and* $y \in \partial K$,

where

$$g \in C_{rd}[\mathbb{T} \times R_+, R_+] \ \text{and} \ u^{\Delta}(t) = g(t, u(t)),\ u(t_0) = 0,$$

has only the trivial solution.

Then $v(t_0) \leq w(t_0)$ *implies* $v(t) \leq w(t)$, $t \in \mathbb{T}$.

Proof: We apply the induction principle to the statement

$$A(t) : v(t) \leq w(t), t \in \mathbb{T}. \tag{4.3.5}$$

(I) $A(t_0)$ is true since $v(t_0) \leq w(t_0)$.

(II) Let t be right-scattered and $A(t)$ be true. We need to show that $A(\sigma(t))$ is true. In view of definition of derivative and property (v) of Theorem 1.2.2, we have,

$$v(\sigma(t)) - w(\sigma(t)) = [\, v^{\Delta}(t) - w^{\Delta}(t)\,]\mu^*(t) + v(t) - w(t),$$

which, because of (4.3.4) and the fact $f(t,x)\mu^*(t) + x$ is non decreasing in x relative to K, yields

$$v(\sigma(t)) - w(\sigma(t)) \leq [\, f(t, v(t)) - f(t, w(t))]\mu^*(t) + v(t) - w(t) \leq 0,$$

proving $A(\sigma(t))$ is valid.

(III) Let t be right dense and N be a neighborhood of t. Suppose that $A(t)$ is true. We have to show that $A(s)$ is true for $s \geq t$, $s \in N$. By Theorem 4.4.3 in [34], it follows that $v(s) \leq w(s)$, $s \geq t$, $s \in N$ and hence $A(s)$ is satisfied.

(IV) Let t be left dense such that $A(s)$ is true for $s < t$. We need to prove that $A(t)$ is true. By rd-continuity of v and w, it follows that

$$v(t) = \lim_{s \to t^-} v(s) \leq \lim_{s \to t^-} w(s) = w(t).$$

Hence $A(t)$ is valid. Thus by induction principle that is, Theorem 1.1.1, we conclude that

$$v(t) \leq w(t),\ t \in \mathbb{T},$$

and the proof is complete.

Remark: We note that K is not assumed to have nonempty interior in Theorem 4.3.3. However, we have imposed the condition that K be a distance set which is a weaker assumption. We observe that every closed convex set in a reflexive Banach space is a distance set and hence when we deal with R^n, this assumption is automatically satisfied. See for details [35].

Consider the IVP for the autonomous dynamic system on time scale $\mathbb{T}$,

$$u^{\Delta} = g(u),\ u(0) = u_0, \tag{4.3.6}$$

relative to which we shall first prove the following result that we need for our discussion.

Theorem 4.3.4: *Let $g \in C[K, E]$, $g(u)$ is quasimonotone nondecreasing in u, $g(u)\mu^*(t) + u$ is nondecreasing in u relative to K, and*

$$[u - v, g(u) - g(v)]_+ \leq L\,|u - v|,\ u, v \in K,\ L \geq 0. \tag{4.3.7}$$

Then there exists a unique solution $u(t, u_0)$ *of* (4.3.7) *for* $t \in \mathbb{T}$ *for every* $u_0 \in K$. *In addition, if* $T(t)u_0 = u(t, u_0)$, *then* $\{T(t)\}$ *is a nonlinear semigroup.*

Proof: We apply Theorem 1.1.1 to the statement

$A(t)$: the IVP $u^{\Delta} = g(u)$, $u(0) = u_0$, has the unique solution $u_{\tau}(t)$ on $[0, \tau]$, $t \in \mathbb{T}$ and $\tau \in \mathbb{T}$.

(I) $A(0)$ is true trivially.

(II) Let τ be right-scattered and $A(\tau)$ be true. Then define the mapping $u_{\sigma(\tau)}(t)$ on $[0, \sigma(\tau)]$ such that

$$u_{\sigma(\tau)}(t) = \begin{cases} u_{\tau}(t), & t \in [0, \tau], \\ u_{\tau}(\tau) + g(u_{\tau}(\tau))\mu^*(\tau), & t = \sigma(\tau). \end{cases}$$

This function is rd-continuous and is the unique solution of (4.3.6) on $[0, \sigma(\tau)]$. Hence $A(\sigma(\tau))$ is true.

(III) Let τ be right-dense and N_{τ} be a neighborhood of τ. Assume $A(\tau)$ is true. we are required to show that $A(s)$ is valid for $s \in N_{\tau}$, $s > \tau$. By local existence and uniqueness there exists a unique solution $u_s(t)$ satisfying

$$u_s^{\Delta}(t) = g(u_s(t)), \quad t \in [\tau, s],\ s \in N_{\tau},$$

$$u_s(\tau) = u_\tau(\tau).$$

Then the mapping defined by

$$u_s(t) = \begin{cases} u_\tau(t), & t \in [0, \tau], \\ u_s(t), & \tau \le t \le s, s \in N_\tau, \end{cases}$$

is the solution of (4.3.6) on $[0, s]$, $s \ge \tau$, which proves $A(s)$ is satisfied.

(IV) Let τ be left-dense such that $A(s)$ is true for all $s < \tau$. We need to prove that $A(\tau)$ is also true. For any $s < \tau$, the IVP (4.3.6) has the unique solution $u_s(t)$ on $[0, s]$ defined by

$$u_s(t) = u_0 + \int_0^t g[u_s(\sigma)]\, d\sigma \;, t \in [0, s],$$

which implies that $A(\tau)$ is true. Thus by Theorem 4.3.3, the IVP (4.3.6) has the unique solution $u(t, u_0)$ for all $t \in \mathbb{T}$. It is now easy to see that if $T(t)u_0 = u(t, u_0)$, then $\{T(t)\}$ satisfies the properties of nonlinear semigroup (see Theorem 2.8.1 in [35]). Hence the proof is complete.

Having Theorem 4.3.3 and 4.3.4 at our disposal, it is now easy to prove the following main result, as before.

Theorem 4.3.5: *Suppose that*

(a) $g \in C[E, E]$, $g(u)$ *is quasimonotone nondecreasing in* u, $g(u)\mu^*(t) + u$ *is non- decreasing in* u *relative to* K, *and for* $\alpha_0 \leq u$, $v \leq \beta_0$, $[u - v, g(u) - g(v)]_+ \leq L\,|u - v|$, $u, v \in K$, $L \geq 0$;

(b) $0 \leq g(\alpha_0)$, $g(\beta_0) \leq 0$ *and* $\alpha_0 \leq \beta_0$;

(c) g *is compact,* K *is normal and a distance set.*

Then (i) $\alpha_0 \leq u(t, \alpha_0)$, $u(t, \beta_0) \leq \beta_0$; $t \in \mathbb{T}$;

(ii) $u(t, u_0) \leq u(t, v_0)$, $t \in \mathbb{T}$ *whenever* $\alpha_0 \leq u_0 \leq v_0 \leq \beta_0$;

(iii) $u(t, u_0)$ *is nondecreasing and* $u(t, \beta_0)$ *is nonincreasing in* $t \in \mathbb{T}$;

(iv) $\lim_{t \to \infty} u(t, \alpha_0) = \underline{u}$, $\lim_{t \to \infty} u(t, \beta_0) = \overline{u}$ *exists and satisfy* $g(\underline{u}) = g(\overline{u}) = 0$.

Proof: We note that by Theorem 4.4.2 in [35], $g(u)$ quasimonotone nondecreasing in u is equivalent to $\lim_{h \to 0^+} \frac{1}{h} d[u + hg(u), K] = 0$, $u \in K$. Hence by Theorem 4.3.4 it follows that the IVP (4.3.6) posses the unique solution $u(t, u_0)$ which exists for $t \in \mathbb{T}$ for every $u_0 \in K$. Now since $0 \leq g(\alpha_0)$ and $u^{\Delta}(t, \alpha_0) = g(u(t, \alpha_0))$, we get by Theorem

4.3.3 arguing as in Theorem 4.4.3 in [34] , $\alpha_0 \le u(t, \alpha_0), t \in \mathbb{T}$. Similarly, by a repeated application of Theorem 4.3.3, we obtain $u(t, \beta_0) \le \beta_0$ and $u(t, u_0) \le u(t, v_0), t \in \mathbb{T}$ whenever $\alpha_0 \le u_0 \le v_0 \le \beta_0$.

This proves (i) and (ii). Since g is autonomous, the solutions of (4.3.6) are unique for $t \in \mathbb{T}$ and satisfy the properties of nonlinear semigroup, using (i) and (ii), it is easy to show that (iii) is satisfied. As a consequence, assumption (c) implies that (iv) holds and the proof is complete.

We remark that if in Theorem 4.3.5 $g(u) = S(u) - u$, then $\underline{u}$, $\overline{u}$ are the fixed points of S. Also, in the present framework, we need to utilize the difference operator to represent difference equations and therefore care must be taken when deducing results corresponding to difference equations.

Let us next consider the IVP on time scale $\mathbb{T}$,

$$x^{\Delta} = H(x), x(0) = x_0, \tag{4.3.8}$$

where $H \in C[E, E]$. We can deduce the following result in view of Theorem 4.3.5.

Theorem 4.3.6: *Assume that*

(i) *there exists a function* $F \in C[E \times E, E]$ *such that* $F(x, x) = H(x)$;

(ii) $F(x, y)$ *is quasimonotone nondecreasing in* $x, F(x, y)\mu^*(t) + x$ *is nondecreasing in* x *and monotone nonincreasing in* y *relative to the cone* K, *and for* $x_1, x_2 \in K$,

$$[x_1 - x_2, F(x_1, y) - F(x_2, y)]_+ \leq L\,|x_1 - x_2|, \;L \geq 0;$$

(iii) $0 \leq F(x_0, y_0), F(y_0, x_0) \leq 0$ *and* $x_0 \leq y_0$.

(iv) F *is compact,* K *is normal and a distance set.*

Then

(a) $x_0 \leq x(t, x_0) \leq x(t, y_0) \leq \beta_0, t \in \mathbb{T}$;

(b) $x(t, x_0)$ *is nondecreasing in t,* $x(t, y_0)$ *is nonincreasing in t,* $\lim_{t \to \infty} x(t, x_0) = \overline{x}\,,\ \lim_{t \to \infty} x(t, y_0) = \overline{y}$ *exist;*

(c) *if* $z(t, z_0)$ *is the solution of* (4.3.8) *with* $x_0 \leq z_0 \leq y_0$, *then* $x(t, x_0) \leq z(t, z_0) \leq x(t, y_0)$, $t \in \mathbb{T}$ *and if* $\overline{z} = \lim_{t \to \infty} z(t, z_0)$ *exists then* $\overline{x} \leq \overline{z} \leq \overline{y}$

Proof: Define the function $g \in C[Y, Y], Y = E \times E$, by

$$g\begin{pmatrix} x \\ y \end{pmatrix} = \begin{pmatrix} F(x, y) \\ F(y, x) \end{pmatrix}$$

and introduce the ordering in $K_Y = K \times \{-K\}$ by

$$\begin{pmatrix} x \\ y \end{pmatrix} \le \begin{pmatrix} x_1 \\ y_1 \end{pmatrix} \text{ iff } x \le x_1 \text{ and } y \ge y_1, \tag{4.3.9}$$

or in other words, if $u = \begin{pmatrix} x \\ y \end{pmatrix}$, $v = \begin{pmatrix} x_1 \\ y_1 \end{pmatrix}$, $u, v \in Y$, $u \le v$ iff (4.3.9) holds.

Define the reflection operator $P: Y \to Y$ by

$$P\begin{pmatrix} x \\ y \end{pmatrix} = \begin{pmatrix} y \\ x \end{pmatrix}.$$

We note that g satisfies the assumptions of Theorem 4.3.5 when we let $\alpha_0 = \begin{pmatrix} x_0 \\ y_0 \end{pmatrix}$, $\beta_0 = P\alpha_0$, $u = \begin{pmatrix} x \\ y \end{pmatrix}$, $T(t)\alpha_0 = u(t, \alpha_0) = \begin{pmatrix} x(t, x_0) \\ x(t, y_0) \end{pmatrix}$. As a result, the conclusions (a) and (b) follow by Theorem 4.3.5. Since $\beta_0 = P\alpha_0$, $u(t, \beta_0) = Pu(t, \alpha_0)$, we have $P\underline{u} = \overline{u}$ where $\underline{u} = \begin{pmatrix} \overline{x} \\ \overline{y} \end{pmatrix}$. To prove (c), we let $u_0 = \begin{pmatrix} z_0 \\ z_0 \end{pmatrix}$ so that $\alpha_0 \le u_0 \le \beta_0$, which implies from (ii) of Theorem 4.3.5 that $u(t, \alpha_0) \le u(t, u_0) \le u(t, \beta_0)$, $t \in \mathbb{T}$ proving (c). Hence the proof of the Theorem 4.3.6 is complete.

4.4 Invariant Manifolds

Let us consider a dynamic equation of the form

$$u^{\Delta} = A^{-}(t)u + r^{-}(t,u,v,w) \; v^{\Delta} = A^{o}(t)v + r^{o}(t,u,v,w) \qquad (4.4.1)$$

$$w^{\Delta} = A^{+}(t)w + r^{+}(t,u,v,w)$$

where u, v and w belong to Banach spaces U, V and W, respectively. For notational reasons, we introduce the symbols

$$\overset{o}{A^{+}}(t) := \begin{pmatrix} A^{o}(t) & 0 \\ 0 & A^{+}(t) \end{pmatrix},$$

$$\overset{o}{r^{+}}(t,u,v,w) := \begin{pmatrix} r^{o}(t,u,v,w) \\ r^{+}(t,u,v,w) \end{pmatrix}$$

$$A^{\bar{o}}(t) := \begin{pmatrix} A^{-}(t) & 0 \\ 0 & A^{o}(t) \end{pmatrix},$$

$$r^{\bar{o}}(t,u,v,w) := \begin{pmatrix} r^{-}(t,u,v,w) \\ r^{o}(t,u,v,w) \end{pmatrix}$$

$$A(t) := \begin{pmatrix} A^{-}(t) & 0 & 0 \\ 0 & A^{o}(t) & 0 \\ 0 & 0 & A^{+}(t) \end{pmatrix},$$

$$r(t,u,v,w) := \begin{pmatrix} r^{-}(t,u,v,w) \\ r^{o}(t,u,v,w) \\ r^{+}(t,u,v,w) \end{pmatrix}$$

and the above abbreviations

$$z := \begin{pmatrix} v \\ w \end{pmatrix} \in Z := V \times W, \quad y := \begin{pmatrix} u \\ v \end{pmatrix} \in Y := U \times V,$$

$$x := \begin{pmatrix} u \\ v \\ w \end{pmatrix} \in X := U \times V \times W .$$

Let us suppose that the system (4.4.1) satisfies the following three hypotheses,

(HI) *Exponential dichotomies*

The transition functions Φ_-, $\Phi_{\overset{o}{+}}$, $\Phi_{\bar{o}}$, Φ_+ of the equations

$$u^\Delta = A^-(t)u\,, \; z^\Delta = A^{\overset{o}{+}}(t)z\,, \; y^\Delta = A^{\overset{o}{-}}(t)y\,, \; w^\Delta = A^+(t)w\,,$$

respectively, satisfy estimates of the form

$$|\Phi_-(t,s)| \le K\,e_\alpha(t,s) \quad \text{for all } t \ge s,$$

$$|\Phi_{\underset{o}{+}}(t,s)| \le L\,e_\beta(t,s) \quad \text{for all } t \ge s,$$

$$|\Phi_{\underline{o}}(t,s)| \le M\,e_\gamma(t,s) \quad \text{for all } t \ge s,$$

$$|\Phi_+(t,s)| \le N\,e_\delta(t,s) \quad \text{for all } t \le s,$$

where K, L, M, N are constants ≥ 1 and α , β ,γ , δ are constants greater than $-\frac{1}{\mu^*(t)} < 0$ for all $t \in \mathbb{T}$.

In case the v -coordinate is present (i.e. $V \neq \{0\}$) it can be shown that $\beta \leq \gamma$.

$(H\,II)$ *Lipschitz conditions*

There exist positive constants $k_{11}, ...m_{22}$ such that

$$\begin{pmatrix} |r^-(t,u_1,z_1) - r^-(t,u_2,z_2)| \\ |\overset{o}{r}{}^+(t,u_1,z_1) - \overset{o}{r}{}^+(t,u_2,z_2)| \end{pmatrix} \leq \begin{pmatrix} k_{11} & k_{12} \\ k_{21} & k_{22} \end{pmatrix} \begin{pmatrix} |u_1 - u_2| \\ |z_1 - z_2| \end{pmatrix}$$

for all $t \in \mathbb{T}$, $y_1, y_2 \in Y$, $w_1, w_2 \in W$.

The inequality symbol " $\leq$ " is to be understood componentwise as usual.

$(H\,III)$ *Spectral conditions*

There exists constants \$ and $\mathbb{Y}$ such that

$$\alpha' : = \$ - (\alpha + K\,k_{11}) > 0, \quad \gamma' : = \mathbb{Y} - (\gamma + M\,m_{11}) > 0$$

$$\beta' : = (\beta - K\,k_{22}) - \$ > 0, \quad \delta' : = (\delta + N\,m_{22}) - \mathbb{Y} > 0$$

$$\theta' : = 1 - \frac{KL.k_{12}k_{21}}{\alpha'\beta'} > 0, \quad \Omega : = 1 - \frac{MN.m_{12}m_{21}}{\gamma'\delta'} > 0$$

$$\Pi : = 1 - \frac{KLMN.k_{12}m_{21}}{\theta\alpha'.\Omega\delta'} > 0$$

Theorem 4.4.1: *Consider system* (4.4.1) *satisfying the assumptions* $(H\,I)$, $(H\,II)$, $(H\,III)$. *Suppose in addition there exist two continuously differentiable functions* $f = (f^-, f^o, f^+) : \mathbb{T} \to X$, $g = (g^-, g^o, g^+) : \mathbb{T} \to X$ *such that the following estimate hold true:*

$$(H\,IV) \quad \limsup_{t \to -\infty} \left|f^{\Delta}(t) - A(t)f(t) - r(t, f(t))\right| e_{\$}(\phi_o, t) < \infty$$

$$\limsup_{t \to -\infty} \left|g^{\Delta}(t) - A(t)g(t) - r(t, g(t))\right| e_{\text{Y}}(\phi_o, t) < \infty.$$

Then for each $\eta \in V$ there exists exactly one solution $(u(t), v(t), w(t))$ of (4.4.1) satisfying the set of conditions

$$\limsup_{t \to -\infty} \left|u(t) - f(t)\right| e_{\$}(\phi_o, t) < \infty, \tag{4.4.2}$$

$$v(\tau) = \eta, \tag{4.4.3}$$

and

$$\limsup_{t \to -\infty} \left|w(t) - g(t)^+\right| e_{\text{Y}}(\phi_o, t) < \infty. \tag{4.4.4}$$

Now several remarks are in order.

Remarks: 1. The solution whose existence and uniqueness has been stated in the theorem can be shown to depend continuously on t, τ, $\eta, f(.)$, and $g(.)$. It will be denoted by

$$x^*(t; \tau, \eta, f(.), g(.)).$$

2. The notations $A^-(t), A^o(t), A^+(t)$ are chosen to recall the classical case of autonomous ordinary differential equations which has been generalized here in various directions. In this case the constant matrices A^-, A^o, A^+ have eigenvalues whose real parts are negative, zero or positive, respectively.

3. Even in case of differential equations with time independent operators A^-, A^o, A^+, Theorem 4.4.1 only requires that the respective spectra can be separated in the complex plane by vertical strips with a certain width depending on the various constants $K, L, .., \gamma, \delta$, $k_{11}, ..., m_{22}$.Two facts need emphasis. First, the spectrum of the center part A^o is not required to be confined to a vertical straight line in C but rather it may contain points with different real parts. Second, the location of the imaginary axis relative to the spectra of A^-, A^o, A^+ doesn't play any role in Theorem 4.4.1.

4. If the time scale underlying the dynamic system 4.4.1 with time independent operators A^-, A^o, A^+ is hZ for some fixed $h > 0$, then our dichotomy assumptions are met if the respective spectra can be separated from each another by concentric rings in the complex plane centered at the real point $-1/h$. Notice that in case $h = 1$ the center of these rings is -1 and note the coordinate origin as it might be expected. This is due to the fact that in our theory the left-hand side $x^{\Delta}(t)$ of the dynamic equations is the difference $x(t+1) - x(t)$ rather than the shift $x(t+1)$.

5. In view of invariant manifolds Theorem 4.4.1 says the following. For each $(\tau, \eta) \in \mathbb{T} \times V$ generates a set of initial points for a certain class of solutions. That this set forms an invariant manifold in $\mathbb{T} \times X$ with far reaching properties will be described in the next theorem. we write equations (4.4.1) in the form

$$x^{\Delta} = A(t)\,x + r(t, x) \tag{4.4.5}$$

and suppose again that it satisfies the assumptions $(H\,I)$, $(H\,II)$, $(H\,III)$. In addition, we consider two equations of the form

$$x^{\Delta} = A(t)\,x + \hat{r}(t, x) \tag{4.4.6}$$

$$x^{\Delta} = A(t)\,x + \widetilde{r}(t,x) \tag{4.4.7}$$

We denote by $x(t;t_0,x_0), \widehat{x}(t;t_0,x_0), \widetilde{x}\ (t;t_0,x_0)$ the respective solutions of (4.4.5), (4.4.6), (4.4.7) with initial value x_0 at $t = t_0$. The requirement on the equations (4.4.6) and (4.4.7) is that for any $t_1, t_2 \in \mathbb{T}$, $x_1, x_2 \in X$ the couple of functions $\widehat{x}(.;t_0,x_0)$, $\widetilde{x}\ (.;t_0,x_0)$ satisfies the hypothesis $(H\ IV)$ above.

For a concise formulation of our theorem on invariant manifolds we introduce the mapping

$$H : \mathbb{T} \times V \times X \times X \to X$$

$$(t,\eta,x_1,x_{2)} \to x^*(t;t,\eta,\widehat{x}(.;t,x_1),\widetilde{x}\ (.;t,x_2)),$$

which can be shown to be continuous.

Theorem 4.4.2: *Suppose the assumption preceding this theorem are fulfilled. Suppose, in addition, that* $f = (f^-, f^o, f^+) : \mathbb{T} \to X$ *and* $g = (g^-, g^o, g^+) : \mathbb{T} \to X$ *are any solutions of* (4.4.6) *and* (4.4.7) *respectively, and that* (τ, η) *is any point in* $\mathbb{T} \times V$. *Then the following is true.*

(*a*) *If* $v(t)$ *is the solution of the initial value problem*

$$v^{\Delta} = A^{o}(t)v + r^{o}(t, H(t, v, f(t), g(t))), \ v(\tau) = \eta, \qquad (4.4.8)$$

then $H(t, v, f(t), g(t))$ *is the* (*by Theorem* 4.4.1 *unique*) *solution of system* (4.4.1) *with* (4.4.2), (4.4.3),(4.4.4).

(*b*) *If* $(u(t), v(t), w(t))$ *is the* (*by Theorem* 4.4.1 *unique*) *solution of system* (4.4.1) *with* (4.4.2), (4.4.3),(4.4.4) *then* $v(t)$ *is the solution of the initial value problem* (4.4.8).

(*c*) *The set*

$$M(\$,\mathbb{Y}) : = \{(t, H(t, v, f(t), g(t))) : (t, v) \subseteq \mathbb{T} \times V\} \in \mathbb{T} \times X$$

is invariant in the following sense

$$(\tau, x_0) \in M(\$,\mathbb{Y}) \Rightarrow (t, x(t; \tau, x_0)) \in M(\$,\mathbb{Y}) \text{ for all } t \in \mathbb{T}.$$

(*d*) $M(\$,\mathbb{Y})$ *contains the trajectories of exactly those solutions* $(u(t), v(t), w(t))$ *of system* (4.4.1) *for which the two conditions*

$$\limsup_{t \to -\infty} \left|u(t) - f^{-}(t)\right| e_{\$}(\phi_o, t) < \infty,$$

$$\limsup_{t \to -\infty} \left|w(t) - g(t)^{+}\right| e_{\mathbb{Y}}(\phi_o, t) < \infty.$$

hold true.

Remarks: The manifold $M(\$,\mathbb{Y})$ as defined above is called $\$ - \mathbb{Y}$-*center manifold* for $f(.)$ and $g(.)$. The case where $f(.)$ and $g(.)$ are the trivial solutions of (4.4.6) and (4.4.7) respectively, is of particular importance. If in this case $\$,\mathbb{Y}$ may be chosen as $\$ < 0 < \mathbb{Y}$ then $M(\$,\mathbb{Y})$ is a classical center manifold for system (4.4.1).

Theorem 4.4.2 in its full generality is needed for the proof of our final theorem on topological equivalence.

Theorem 4.4.3: *Consider system* (4.4.1) *obeying* $(H\,I)$, $(H\,II)$ *and* $(H\,III)$. *In addition, suppose that*

$$\limsup_{t \to -\infty} |r(t,0)| e_{\$}(\phi_o, t) < \infty,$$

$$\limsup_{t \to \infty} |r(t,0)| e_{\mathbb{Y}}(\phi_o, t) < \infty$$

Then there exist continuous mappings

$$\underline{h} : \mathbb{T} \times U \times V \times W \to U \times V \times W$$

$$h : \mathbb{T} \times U \times V \times W \to U \times V \times W$$

with the following properties:

(a) *If* $(u(t), v(t), w(t))$ *is a solution of* (4.4.1), *then* $(u(t), v(t), w(t)) := \underline{h}\,(u(t), v(t), w(t))$ *is a solution of*

$$u^{\Delta} = A^{-}(t)u$$

$$v^{\Delta} = A^{o}(t)v + r^{o}(t, H(t, v, 0, 0) \tag{4.4.9}$$

$$w^{\Delta} = A^{+}(t)w$$

such that

$$\limsup_{t \to -\infty} \left|\underline{u}\,(t) - u(t)\right| e_{\$}(\phi_o, t) < \infty,$$
$$\limsup_{t \to \infty} \left|\underline{w}\,(t) - w(t)\right| e_{Y}(\phi_o, t) < \infty.$$

(b) *If* $(\underline{u}\,(t), \underline{v}\,(t), \underline{w}\,(t))$ *is a solution of* (4.4.1), *then* $(u(t), v(t), w(t)) := h(\underline{u}\,(t), \underline{v}\,(t), \underline{w}\,(t))$ *is a solution of* (4.4.1) *such that*

$$\limsup_{t \to -\infty} \left|u(t) - \underline{u}\,(t)\right| e_{\$}(\phi_o, t) < \infty,$$
$$\limsup_{t \to \infty} \left|w(t) - \underline{w}\,(t)\right| e_{Y}(\phi_o, t) < \infty.$$

(c) *For each* $t \in \mathbb{T}$ *the mappings* $\underline{h}\,(t, .)$ *an* $h(t, .)$ *are complementary homeomorphisms on* $U \times V \times W$.

The proofs of the results presented in this section are rather technical and lengthy. They can be found in full detail in Hilger [18]. Roughly speaking they are modeled after Palmer [52] and Aulbach [3]. The proofs show that calculus on measure chains as presented in the first chapter is quite well adapted to the needs of a qualitative theory of dynamic equations.

4.5 Practical Stability of Large-Scale Uncertain Dynamic Systems

Consider the following uncertain dynamic system

$$x^{\Delta} = F(t, x, u, \omega) \qquad x(t_0) = x_0 \qquad t_0 \geq 0, \qquad (4.5.1)$$

where $t \in \mathbb{T}$ (in the sequel, we always mean by time interval the intersection of a real interval with the given time scale), $x \in R^n$ is the state, and $u \in R^m$ is the control input. The uncertainty in the system is represented by ω. $F \in C_{rd}\,[\mathbb{T} \times R^n \times R^m \times \Omega, R^n]$ is known and so is the set Ω to which ω belongs. No other information on ω is known.

Suppose that F is smooth enough to guarantee existence, uniqueness, and rd-continuous dependence of solutions $x(t) = x(t, t_0, x_0, u, \omega)$ of (4.5.1). For systems described by (4.5.1) we consider controllers of the form

$$u(t) = p(t,x(t))$$

where $p: \mathbb{T} \times R^n \to R^m$. The goal is to determine the set U of admissible controllers u that ensure the desired stability properties of (4.5.1) for any $\omega \in \Omega$.

Assume that (4.5.1) can be decomposed into N subsystems

$$x^{\mu\Delta} = F^{\mu}(t,x,u,\omega), \qquad x^{\mu}(t_0) = x_0^{\mu} \tag{4.5.2}$$

where $x^{\mu} \in R^{n_\mu}$, $\mu = 1,2, \ldots,N$, and $\sum_{\mu=1}^{N} n_\mu = n$. In addition, also assume that $F^{\mu}(t,x,u,\omega)$ can be decomposed as follows

$$F^{\mu}(t,x,u,\omega) = f^{\mu}(t,x) + B^{\mu}(t,x)\, g^{\mu}(t,x,u,\omega) \tag{4.5.3}$$

where $B^{\mu}(t,x) \in C_{rd}[\mathbb{T} \times R^n, R^{m_\mu n_\mu}]$ is an $n_\mu \text{x}\, m_\mu$ matrix, $f^{\mu} \in C_{rd}[\mathbb{T} \times R^n, R^{n_\mu}]$, $g^{\mu} \in C_{rd}[\mathbb{T} \times R^n \times R^m \times \Omega, R^{m_\mu}]$ and $\sum_{\mu=1}^{N} m_\mu = m$. Systems of the form (4.5.3) are said to have matched uncertainties in the sense that the uncertainty ω and the control u enter the system via the same matrix $B^{\mu}(t,x)$.

First, we prove a theorem which specifies admissible control functions $u(t) = p(t,x(t))$ corresponding to the practical stability of (4.5.1).

Theorem 4.5.1: *Assume for* $1 \leq \mu \leq N$

(A1) $0 < \lambda < A$

(A2) $h_0,\ h \in \Gamma$, *and* h_0 *is uniformly finer than* h, *i.e.*, $h(t,x) \leq \phi(h_0(t,x))$, $\phi \in \mathcal{K}$ *whenever* $h_0(t,x) < \lambda$;

(A3) *there exist* $V^\mu \in C^1_{rd}[\mathbb{T} \times R^n,\ R_+\ ]$, $V^\mu(t,u)$ *is locally Lipschitzian in u, and for* $(t,\ x) \in S(h,A) = \{(t,x) \in \mathbb{T} \times R^n\colon h(t,x) < A\}$, $a, b \in \mathcal{K}$

$$b(h(t,x)) \leq \sum_{\mu=1}^{N} V^\mu(t,\ x) \qquad \text{if } h(t,x) < A$$

$$\sum_{\mu=1}^{N} V^\mu(t,x) \leq a(h_0(t,x)) \qquad \text{if } h_0(t,x) < \lambda, \tag{4.5.4}$$

(A4) *for* $\omega \in \Omega$ *and* $(t,x) \in S(h,A)$

$$V_f^{\mu\Delta}(t,x,\omega) = V_t^{\mu\Delta}(t,x) + V_x^{\mu\Delta}(t,x).\, f^\mu(t,x,\omega)$$

$$\leq G^\mu(t, V(t,x)) \tag{4.5.5}$$

where $G^\mu \in C_{rd}[\mathbb{T} \times R_+, R_+]$, $G_i^\mu(t,u)\mu^*(t) + u_i$ *is nondecreasing in u for each* $t \in \mathbb{T}$;

(A5) $\phi(\lambda) < A$ *and* $a(\lambda) < b(A)$ *hold;*

(A6) *for* $(t,x) \in S(h,A)$ *and* $\omega \in \Omega$, *there exists*

$$p(t,x) = (p^1(t,x)\ p^2(t,x) \ldots p^q(t,x)),\quad p^\mu \in C_{rd}[\mathbb{T} \times R^n, R^{m_\mu}],$$

$\mu = 1, 2, ..., N$, *such that the following conditions are satisfied*

$$(a) \quad p^{\mu T}(t,x) g^{\mu}(t,x,p(t,x),\omega) \geq$$

$$(-\beta_1^{\mu}(t,x,\omega) + \beta_2^{\mu}(t,x,\omega) W_0^{\mu}(\|p^{\mu}(t,x)\|) \ \|p^{\mu}(t,x)\| \tag{4.5.6}$$

$$\beta_1^{\mu}(t,x,\omega) \leq \beta_2^{\mu}(t,x,\omega)\, \rho^{\mu}(t,x) \tag{4.5.7}$$

$$\beta_1^{\mu}(t,x,\omega) \leq \kappa^{\mu}(t,x) \tag{4.5.8}$$

$$\beta_1^{\mu},\ \beta_2^{\mu} \in C_{rd}[\mathbb{T} \times R^n \times \Omega,\ R_+],\ \ W_0^{\mu} \in C[R_+,\ R_+],$$

$$\text{and } \kappa^{\mu},\ \rho^{\mu} \in C[R \times R^n,\ R_+];$$

(b) *for* $(t,x) \in S(h,A)$, $\alpha^{\mu}(t,x) \in R^{m_{\mu}}$,

$$\|\alpha^{\mu}(t,x)\| \ p^{\mu}(t,x) = -\|p^{\mu}(t,x)\| \alpha^{\mu}(t,x) \tag{4.5.9}$$

where $\alpha(t,x) \equiv B^{\mu}(t,x)^{T} V_x^{\mu}(t,x)^{T}$,

(c) $\eta^{\mu}(t,x) \equiv \kappa^{\mu}(t,x)\, \alpha^{\mu}(t,x)$, $\|\eta^{\mu}(t,x)\| > 0$, *implies*

$$W_0^{\mu}(\|p(t,x)\|) \geq \rho^{\mu}(t,x)[1 - \|\eta^{\mu}(t,x)\|^{-1} W^{\mu}(t,x)] \tag{4.5.10}$$

where $W^{\mu} \in C_{rd}[\mathbb{T} \times R_+, R_+]$, $W^{\mu}(t,x) \leq W^{*\mu}(t, V(t,x))$ *for* $(t,x) \in S(h,A)$ *and* $W^{*\mu}(t,u)$ *nondecreasing in u,*

(A7) the system

$$u' = G(t,u) + W^{*}(t,u)\,, \quad u(t_0) = u_0 > 0 \tag{4.5.11}$$

where $W^{*}(t,u) \in C_{rd}[\mathbb{T} \times R_+^N,\ R_+^N]$, *is practically stable with respect* *to* $(a(\lambda), b(A))$ *then system* (4.5.1) *is* (h_0,h)*-practically stable.*

Proof: From equations (4.5.9), (4.5.6), (4.5.10), and then (4.5.7), (4.5.8) we have, for i=1, 2, ..., q

$$\alpha^\mu(t,x)g^\mu(t,x,p,\omega) = -\frac{\|\alpha^\mu(t,x)\|}{\|p^\mu(t,x)\|}\, p^\mu\ (t,x)\ g^\mu(t,p,u,w)$$

$$\leq \|\alpha^\mu(t,x)\| \left(\beta_1^\mu(t,x,\omega) - \beta_2^\mu(t,x,\omega)\ W_0^\mu(\|p^\mu(t,x)\|)\right)$$

$$\leq \|\alpha^\mu(t,x)\| \left(\beta_1^\mu(t,x,\omega) - \beta_2^\mu(t,x,\omega)\rho^\mu(t,x)\left(1 - \frac{W^\mu(t,x)}{\|\eta^\mu(t,x)\|}\right)\right)$$

$$\leq \frac{\|\alpha^\mu(t,x)\|\ \beta_1^\mu(t,x,\omega)}{\|\eta^\mu(t,x)\|}\ W^\mu(t,x)$$

$$\leq \frac{\beta_1^\mu(t,x,\omega)}{\kappa^\mu(t,x)}\ W^\mu(t,x) \leq W^\mu(t,x) \tag{4.5.12}$$

Now consider

$$V_F^{\mu\Delta}(t,x(t),\omega) = V_t^{\mu\Delta}(t,x(t)) + V_x^{\mu\Delta}(t,x(t))\, f^\mu(t,x(t),\omega)$$

$$+\ V_x^{\mu\Delta}(t,x(t))\, B^\mu(t,x(t))\ g^\mu(t,x(t),p(t,x(t)),\omega)$$

Using (4.5.5), (4.5.12) we have, for $\mu = 1, 2, ..., N$,

$$V_F^{\mu\Delta}(t,x(t),\omega) \leq G^\mu(t,\ V^\mu(t,\ x(t)) + W^\mu(t,\ x(t))$$

$$\leq G^\mu(t,\ V^\mu(t,\ x(t)) + W^{*\mu}(t,\ V(t,x(t))$$

$$\equiv F^\mu(t,\ V(t,\ x(t))).$$

Since all the conditions of Theorem 3.6.1 are satisfied, (4.5.1) is (h_0,h)-practically stable.

Similarly, we can prove (h_0,h)-practical asymptotic stability of (4.5.1) using Theorem 4.4.3.

Theorem 4.5.2: *Assume that conditions (A1)-(A5) of Theorem* 4.5.1 *are satisfied with the following modifications:*

(i) *equation* (4.5.5) *is replaced by*

$$V_f^{j\Delta}(t,x,\omega) \le -\gamma(h_0(t,x)) \qquad 1 \le j \le N \tag{4.5.13}$$

$$V_f^{i\Delta}(t,x,\omega) \le 0 \qquad 1 \le i \le N,\ i \ne j$$

where $\gamma \in \mathcal{K}$, *and* $h_0(t,x)$ *satisfies the same condition as in Theorem* (4.4.2),

(ii) *equation* (4.5.10) *is replaced by*

$$W_0^{\mu}(\|p^{\mu}(t,x)\|) \ge \rho^{\mu}(t,x)\,[1 - \|\eta^{\mu}(t,x)\|^{-1}\,\epsilon^{\mu}(t)] \tag{4.5.14}$$

where $\epsilon^{\mu} \in L^1$ *and* $0 \le \epsilon^{\mu}(t,x) \le \eta^{\mu}(t,x)$ *in* $S(h,A)$.

Then system (4.5.1) *is* (h_0,h)*-practically asymptotically stable.*

Proof: Following the same steps as in the proof of Theorem 4.5.1 we can show that

$$V_F^{j\Delta}(t,x,\omega) \leq -\gamma(h(t,x)) + \epsilon_j(t) \qquad 1 \leq j \leq N \qquad (4.5.15)$$

$$V_F^{i\Delta}(t,x,\omega) \leq \epsilon_i(t) \qquad 1 \leq i \leq N,\ i \neq j$$

Hence, by Theorem 4.4.3 we conclude that (4.5.1) is (h_0,h)-practically asymptotically stable.

Example: Consider the following system

$$
\begin{aligned}
x_1^{\Delta} &= -\rho_1 x_1 \\
x_2^{\Delta} &= -\rho_2 x_2 + R^1(t,x,p,\omega) \\
x_3^{\Delta} &= -\rho_3 x_3 + x_5 \qquad\qquad (4.5.16) \\
x_4^{\Delta} &= -\rho_4 x_4 + x_5 \\
x_5^{\Delta} &= -b_3 x_3 - b_4 x_4 - x_5 + R^2(t,x,p,\omega) \\
x_i(t_i) &= x_{i0}, \qquad i = 1,2,3,4,5
\end{aligned}
$$

where $\rho_i > 0$, $i = 1, 2, 3, 4$, b_3, b_4 are constants, $\omega \in \Omega = \{\psi \in R : |\psi| < \omega_0\}$ is an uncertain parameter, $p \in C[\mathbb{T} \times R^5, R^4]$ is the control, and

$$R^1(t,x,p,\omega) = -\,\omega^2 x_1^4\ cost + p^{(1)^T}(t,x).v^{(1)}(t,x)$$

$$R^2(t,x,p,\omega) = -\,\omega^2 (x_3^2 + x_4^2)^2\ sint + p^{(2)^T}(t,x).v^{(2)}(t,x)$$

where $v^{(1)}(t,x) = \omega^2 \begin{bmatrix} x_1^2 \\ 1 \end{bmatrix}$, $\quad v^{(2)}(t,x) = \begin{bmatrix} x_3^2 + x_4^2 \\ 1 \end{bmatrix}$

In this example our goal is to specify an admissible control function that assures a practical asymptotic behavior of the controlled motion despite the presence of the uncertain parameter ω. For this purpose let us decompose the control system (4.5.16) into the following subsystems

$$S_1:\ X_1^{\Delta} = F^{(1)}(t,x) + B^{(1)}(t,x)g^{(1)}(t,x,p,\omega)$$

$$S_2:\ X_2^{\Delta} = F^{(2)}(t,x) + B^{(2)}(t,x)g^{(2)}(t,x,p,\omega)$$

where

$$X_1^T = (x_1\ \ x_2) \qquad X_2^T = (x_1\ \ x_2\ \ x_3)$$

$$F^{(1)}(X_1) = \begin{bmatrix} -\,\rho_1 x_1 \\ -\,\rho_2 x_2 \end{bmatrix} \qquad B^{(1)}(X_1) = \begin{bmatrix} 0 & 0 \\ 0 & x_1^2 \end{bmatrix}$$

$$g^{(1)}(t,x,p,\omega) = \begin{bmatrix} -\,\omega^2 x_1^4 cos\, t + p_1^{(1)}(t,x)\omega^2 x_1^2 \\ \omega^2 p_2^{(1)}(t,x) \end{bmatrix}$$

$$F^{(2)}(X_2) = \begin{bmatrix} -\rho_3 x_3 + x_5 \\ -\rho_4 x_4 + x_5 \\ -b_3 x_3 - b_4 x_4 - x_5 \end{bmatrix} \qquad B^{(2)}(X_2) = \begin{bmatrix} 0 & 0 \\ 0 & 0 \\ 0 & x_3^2 + x_4^2 \end{bmatrix}$$

$$g^{(2)}(t,x,p,\omega) = \begin{bmatrix} -\omega^2(x_3^2 + x_4^2)^2 \cos t + p_1^{(2)}(t,x)\omega^2(x_3^2 + x_4^2) \\ \omega^2 p_2^{(2)}(t,x) \end{bmatrix}$$

Consider the unperturbed and uncontrolled systems

$$S_1^*\colon X_1^\Delta = \begin{bmatrix} -\rho_1 x_1 \\ -\rho_2 x_2 \end{bmatrix} \qquad S_2^*\colon X_2^\Delta = \begin{bmatrix} -\rho_3 x_3 + x_5 \\ -\rho_4 x_4 + x_5 \\ -b_3 x_3 - b_4 x_4 - x_5 \end{bmatrix}$$

for which we define the following Lyapunov functions

$$V_1(X_1) = \tfrac{1}{2}\left(| b_1 | x_1^2 + | b_2 | x_2 V_2(X_2)\right.$$

$$= \tfrac{1}{2}\left(| b_3 | x_3^2 + | b_4 | x_4^2 + x_5^2\right)$$

where the b_i's are constants. Defining

$$V_0(x) = V_1(X_1) + V_2(X_2)\,,$$

$$h(t,x) = \left(\sum_{i=3}^{5} x_i^2\right)^{1/2} \qquad h_0(t,x) = \left(\sum_{i=1}^{5} x_i^2\right)^{1/2}$$

$$\eta_1 = min(1, | b_3 |, | b_4 |), \quad \eta_2 = max(1, | b_i |, i = 1,2,3,4)\,,$$

it is easy to show that, given $0 < \lambda < A$ such that $\frac{\lambda^2}{A^2} < \frac{\eta_1}{\eta_2} < 1$, we have

$$b(h(t,x)) \equiv \frac{1}{2}\,\eta_1 h^2(t,x) \le V_0(t,x) \qquad \text{for } h(t,x) < A$$

$$V_0(t,x) \le \frac{1}{2}\,\eta_2 h_0^2(t,x) \equiv a(h_0(t,x)) \qquad \text{for } h_0(t,x) < \lambda$$

$$a(\lambda) = \frac{1}{2}\,\eta_2\lambda^2 < \frac{1}{2}\,\eta_1 A^2 = b(A).$$

Moreover, for $(t,x) \in S(h,A)$, $\eta_3 \equiv max(1,\rho_3,\rho_4)$, $\eta_4 \equiv max(1, \mid b_3 \mid, \mid b_4 \mid)$ we have

$$h^{\Delta}(t,x) \ge -2\,[\,\mid b_3 \mid x_3^2 + \mid b_4 \mid x_4^2 + x_5^2]^{1/2}$$

$$\ge 2\eta_3\eta_4 A \equiv -M, \qquad M > 0.$$

The derivatives of the Lyapunov functions $V_1(X_1)$ and $V_2(X_2)$ along the solutions of S_1^* and S_2^*, respectively are

$$V_1^{\Delta}(X_1) = -\rho_1 \mid b_1 \mid x_1^2 - \rho_2 \mid b_2 \mid x_2^2 \le 0$$

$$V_2^{\Delta}(X_2) = -\mid b_3 \mid x_3^2 - \mid b_4 \mid x_4^2 - x_5^2$$

$$\le -\eta_1[x_3^2 + x_4^2 + x_5^2] = -\eta_1 h^2\,(t,x).$$

In addition, we have

$$p^{(1)T} g^{(1)}(t,x,p,\omega) = -\omega^2 x_1^4 cos\,t\; p_1^{(1)}(t,x) + \omega^2 x_1^2 \left\|p_1^{(1)}(t,x)\right\|^2$$

$$\ge -2\omega^2 x_1^4\, cos\,t \left\|p_1^{(1)}(t,x)\right\| + \omega^2 x_1^2 \left\|p_1^{(1)}(t,x)\right\|^2$$

Hence, using the notation introduced in Theorem 4.5.1 we have

$$\beta_1^{(1)}(t,x,\omega) = 2\omega^2 x_1^4 \cos t \;\leq\; 2\omega^2 x_1^4, \qquad \beta_2^{(1)}(x,\mathrm{w}) = \omega^2 x_1^2$$

$$\rho^{(1)}(x) = 2x_1^2 \qquad \mathcal{K}^{(1)}(x) = 2\omega_0^2 x_1^4$$

$$\alpha^{(1)}(x) = B^{(1)}(t,x)\mathrm{V}_1^{\Delta\mathrm{T}}(x) = x_2 \mid b_2 \mid \begin{bmatrix} 1 \\ x_1^2 \end{bmatrix}$$

$$\eta^{(1)}(t,x) = \mathcal{K}^{(1)}(x)\alpha^{(1)}(t,x) = 2\omega_0^2 x_1^4 x_2 \mid b_2 \mid \begin{bmatrix} 1 \\ x_1^2 \end{bmatrix}$$

$$\left\|\eta^{(1)}(t,x)\right\| = 2\omega_0^2 x_1^4 \mid b_2 x_2 \mid (x_1^4 + 1)^{1/2}.$$

Similarly, we have

$$p^{(2)^T} g^{(2)}(t,x,p,\omega) =$$

$$- \omega^2(x_3^2 + x_4^2)^2 \sin t \; p_1^{(2)}(t,x) + \omega^2(x_3^2 + x_4^2) \left\|p_1^{(2)}(t,x)\right\|^2$$

$$\geq - 2\omega^2(x_3^2 + x_4^2)^2 \sin t \left\| p_1^{(2)}(t,x)\right\| + \omega^2(x_3^2 + x_4^2) \left\|p_1^{(2)}(t,x)\right\|^2$$

$$\beta_1^{(2)}(t,x,\omega) = 2\omega^2(x_3^2 + x_4^2)^2 \sin t \;\leq\; 2\omega^2(x_3^2 + x_4^2)^2,$$

$$\beta_2^{(2)}(x,\mathrm{w}) = \omega^2(x_3^2 + x_4^2)$$

$$\rho^{(2)}(x) = 2(x_3^2 + x_4^2) \qquad \mathcal{K}^{(1)}(x) = 2\omega_0^2(x_3^2 + x_4^2)^2$$

$$\alpha^{(2)}(X_2) = B^{(2)T}(X_2)V_1^{\Delta T}(X_2) = 2x_5 \begin{bmatrix} 1 \\ x_3^2 + x_4^2 \end{bmatrix}$$

$$\eta^{(2)}(x) = \mathcal{K}^{(1)}(x)\alpha^{(1)}(t,x) = 4\omega_0^2(x_3^2 + x_4^2)^2\, x_5 \begin{bmatrix} 1 \\ x_3^2 + x_4^2 \end{bmatrix}$$

$$\left\|\eta^{(2)}(t,x)\right\| = 4\omega_0^2(x_3^2 + x_4^2)^2 \mid x_5 \mid (x_3^2 + x_4^2)^2 + 1)^{1/2}.$$

To satisfy condition (A5c) of Theorem 4.5.1 let us choose

$$W^{\mu}(t,x) = W^{\mu}(t,x) = e^{-t}.$$

Then, an admissible control function $p^T(t,x) = (p^{(1)}(t,x) \quad p^{(2)}(t,x))$ is given, for $t \geq t_0$, by

$$p^{(1)}(t,x) = \begin{cases} -\eta^{(1)}(t,x)e^t & \left\|\eta^{(1)}(t,x)\right\| \leq e^{-t} \\ \frac{-\eta^{(1)}(t,x)}{\left\|\eta^{(1)}(t,x)\right\|}\, 2x_1^2 & \left\|\eta^{(1)}(t,x)\right\| > e^{-t} \end{cases}$$

$$p^{(2)}(t,x) = \begin{cases} -\eta^{(2)}(t,x)e^t & \left\|\eta^{(2)}(t,x)\right\| \leq e^{-t} \\ \frac{-\eta^{(2)}(t,x)}{\left\|\eta^{(2)}(t,x)\right\|}\, 2(x_3^2 + x_4^2) & \left\|\eta^{(2)}(t,x)\right\| > e^{-t}. \end{cases}$$

4.6 Boundary Value Problems

We shall discuss, in this section, existence and uniqueness of solutions of two point boundary value problems. we shall also study the properties of the Green's matrix.

Consider the linear dynamic system

$$x^{\Delta} = A(t)\,x + b(t),\ x(\tau) = x_0, \tau \in \mathbb{T}^k, \tag{4.6.1}$$

where $A : \mathbb{T}^k \to B(R^n)$ is regressive and rd-continuous, and $b : \mathbb{T}^k \to R^n$ is rd-continuous. We shall suppose that $\mathbb{T}^k = [a, b] \cap \mathbb{T}$. The associated homogeneous linear dynamic system is

$$x^{\Delta} = A(t)\,x,\ x(\tau) = x_0 \tag{4.6.2}$$

We recall that any solution $x(t)$ of (4.6.1) has the form

$$x(t) = \Phi_A(t,\tau)\,x_0 + \int_\tau^t \Phi_A(t,\sigma(s))\,b(s)\ \Delta\, s$$

where $\Phi_A(t,\tau)$ is the fundamental matrix solution of

$$x^{\Delta} = A(t)\,x$$

with $\Phi_A(\tau,\tau) = I$ (Identity matrix). (see Theorem 2.5.1). Also, it is easy to see that a particular solution $\overline{x}\ (t)$ of

$$x^{\Delta} = A(t)\,x + b(t),$$

is of the form $\int_\tau^t \Phi_A(t,\sigma(s))\,b(s)\ \Delta\, s$.

Definition 4.6.1: If $\Phi_A(t)$ is a fundamental matrix solution, then the matrix D defined by

$$D = M\Phi_A(a) + N\Phi_A(b)$$

is called the characteristic matrix for the boundary value problem (4.6.1) satisfying the boundary conditions $Mx(a) + Nx(b) = 0$.

Definition 4.6.2: The number of linearly independent solutions of the boundary value problem is called the index of compatibility of the boundary value problem, If the index of compatibility of the boundary value problem is zero then we say that the boundary value problem is incompatible.

Theorem 4.6.1: *If the boundary value problem*

$$x^{\Delta} = A(t)\,x + b(t)$$

$$Mx(a) + Nx(b) = 0.$$

has a characteristic matrix of rank r then its index of compatibility is $(n - r)$.

Proof: Let $\Phi_A(t)$ be a fundamental matrix of

$$x^{\Delta} = A(t)\,x.$$

Then any solution is given by $\Phi_A(t)C$, where C is a constant n-vector and will be determined uniquely from the boundary conditions. Substituting the general solution of

$$x^{\Delta} = A(t)\,x$$

in the boundary condition matrix we get

$$[\,M\Phi_A(a) + N\Phi_A(b)\,]C = 0,$$

$$DC = 0,$$

since the rank of D is r, and DC has $(n-r)$ linearly independent solutions. Hence, the index of compatibility of the boundary value problem is $(n-r)$. we shall establish the existence and uniqueness solutions to two-point boundary value problems on an in-homogeneous time-scale of a system of the linear dynamic process by utilizing the variation of parameters formula. the solution is established as an integral representation of Green's matrix and the properties of the Green's matrix are also studied. Our main result is stated in the following theorem.

Theorem 4.6.2: *Let* $A : \mathbb{T}^k \to B(R^n)$ *be rd-continuous and regressive and let* $b : \mathbb{T}^k \to R^n$ *be continuous. Then there exists a unique solution of the linear dynamic equation,*

$$x^\Delta = A(t)\, x + b(t)$$

satisfying the general boundary conditions,

$$Mx(a) + Nx(b) = \alpha\,,$$

where M *and* N *are constant square matrices of order* n, *and* $x(t)$ *is given by*

$$x(t) = \Phi_A(t)\, D^{-1}\alpha + \int_a^b G(t, \sigma(s))\, b(s) \,\Delta\, s$$

where $G(t, \sigma(s))$ *is the Green's matrix for the homogeneous linear dynamic boundary value problem.*

Proof: Any solution of the linear dynamic equation

$$x^\Delta = A(t)\, x + b(t)$$

is given by

$$x(t) = \Phi_A(t) C + \int_a^t \Phi_A(t, \sigma(s))\, b(s) \,\Delta\, s$$

Substituting the general form of $x(t)$ in the boundary condition matrix we get,

$$[\, M\Phi_A(a) + N\Phi_A(b)\,]C + N\int_a^b \Phi_A(t, \sigma(s))\, b(s) \,\Delta\, s = \alpha,$$

$$DC = \alpha - N\int_a^b \Phi_A(t,\sigma(s))\, b(s) \,\Delta s,$$

$$C = D^{-1}\alpha - D^{-1}N\int_a^b \Phi_A(t,\sigma(s))\, b(s) \,\Delta s.$$

Hence

$$\begin{aligned}
x(t) &= \Phi_A(t)\, D^{-1}\alpha - \Phi_A(t)\, D^{-1}N\int_a^b \Phi_A(t,\sigma(s))\, b(s) \,\Delta s \\
&\quad + \int_a^b \Phi_A(t,\sigma(s))\, b(s) \,\Delta s \\
&= \Phi_A(t)\, D^{-1}\alpha - \Phi_A(t)\, D^{-1}N\Phi_A(b)\int_a^b \Phi_A^{-1}(\sigma(s))\, b(s) \,\Delta s \\
&\quad + \Phi_A(t)\int_a^b \Phi_A^{-1}(\sigma(s))\, b(s) \,\Delta s
\end{aligned}$$

$$\begin{aligned}
&= \Phi_A(t)\, D^{-1}\alpha - \Phi_A(t)\, D^{-1}N\Phi_A(b)[\int_a^t \Phi_A^{-1}(\sigma(s))\, b(s) \,\Delta s \\
&\quad \int_t^b \Phi_A^{-1}(\sigma(s))\, b(s) \,\Delta s] + \Phi_A(t)\int_a^t \Phi_A^{-1}(\sigma(s))\, b(s) \,\Delta s \\
&= \Phi_A(t)\, D^{-1}\alpha + \Phi_A(t)[I - D^{-1}N\Phi_A(b)]\int_a^t \Phi_A^{-1}(\sigma(s))\, b(s) \,\Delta s \\
&\quad - \Phi_A(t)\, D^{-1}N\Phi_A(b)\int_t^b \Phi_A^{-1}(\sigma(s))\, b(s) \,\Delta s \\
&= \Phi_A(t)\, D^{-1}\alpha + \Phi_A(t)[DD^{-1} - D^{-1}N\Phi_A(b)]\int_a^t \Phi_A^{-1}(\sigma(s))\, b(s) \,\Delta s \\
&\quad - \Phi_A(t)\, D^{-1}N\Phi_A(b)\int_t^b \Phi_A^{-1}(\sigma(s))\, b(s) \,\Delta s \\
&= \Phi_A(t)\, D^{-1}\alpha + \Phi_A(t)D^{-1}[D - N\Phi_A(b)]\int_a^t \Phi_A^{-1}(\sigma(s))\, b(s) \,\Delta s \\
&\quad - \Phi_A(t)\, D^{-1}N\Phi_A(b)\int_t^b \Phi_A^{-1}(\sigma(s))\, b(s) \,\Delta s \\
&= \Phi_A(t)\, D^{-1}\alpha + \Phi_A(t)D^{-1}M\Phi_A(a)\int_a^t \Phi_A^{-1}(\sigma(s))\, b(s) \,\Delta s \\
&\quad - \Phi_A(t)\, D^{-1}N\Phi_A(b)\int_t^b \Phi_A^{-1}(\sigma(s))\, b(s) \,\Delta s
\end{aligned}$$

$$= \Phi_A(t)\, D^{-1}\alpha + \int_a^b G(t,\sigma(s))\, b(s) \,\Delta\, s,$$

where $G(t,\sigma(s))$ is the Green's matrix given by

$$G(t,\sigma(s)) = \begin{cases} \Phi_A(t) D^{-1} M \Phi_A(a)\ \Phi_A^{-1}(\sigma(s)). & a \le s < t \le b \\ -\ \Phi_A(t) D^{-1} N \Phi_A(b)\ \Phi_A^{-1}(\sigma(s)), & a \le t < s \le b, \end{cases}$$

which can be written conveniently as

$$G(t,\sigma(s)) = \begin{cases} \Phi_A(t) D^{-1} M \Phi_A(a,\sigma(s)), & a \le s < t \le b \\ -\ \Phi_A(t) D^{-1} N \Phi_A(b,\sigma(s)), & a \le t < s \le b, \end{cases}$$

Theorem 4.6.3: *The Green's matrix G has the following properties:*

(i) *The components of $G(t,\sigma(s))$ when regarded as a function of t with fixed s have continuous first derivatives on $[a,s)$ and $(s,b]$. at the point $t = s$, G has an upward jump discontinuity, i.e.,*

$$G(s^+,\sigma(s)) - G(s^-,\sigma(s)) = \Phi_A(s,\sigma(s)) = \Phi_A(s)\ \Phi_A^{-1}(\sigma(s)).$$

If $\Phi_A^{-1}(\sigma(s))$ is rd-continuous then G has a finite jump discontinuity of unit magnitude.

(ii) *G has a formal solution of the homogeneous boundary value problem and it fails to be a true solution because of the discontinuity at $t = s$.*

(iii) *G is unique with properties* (i) *and* (ii).

Proof: (i) The Green's matrix G defined in the above theorem may be written as

$$G(t,\sigma(s)) = \begin{cases} \Phi_A(t)\, H_+, & s < t \\ \Phi_A(t)\, H_-, & t < s, \end{cases}$$

where H_+ and H_- are free from t and are given by

$$H_+ = D^{-1} M \Phi_A(a,\sigma(s))$$

and

$$H_- = \; - D^{-1} N \Phi_A(b,\sigma(s)).$$

Consider,

$$H_+ - H_- = D^{-1}M\Phi_A(a,\sigma(s)) + D^{-1}N\Phi_A(b,\sigma(s))$$

$$= D^{-1}M\Phi_A(a)\Phi_A^{-1}(\sigma(s)) + D^{-1}N\Phi_A(b)\Phi_A^{-1}(\sigma(s))$$

$$= D^{-1}[\,M\Phi_A(a) + N\Phi_A(b)]\,\Phi_A^{-1}(\sigma(s))$$

$$= D^{-1}D\,\Phi_A^{-1}(\sigma(s))$$

$$= \Phi_A^{-1}(\sigma(s)).$$

Hence,

$$G(s^+,\sigma(s)) - G(s^-,\sigma(s)) = \Phi_A(s)H_+ - \Phi_A(s)H_-$$

$$= \Phi_A(s)[\,H_+ - H_-]$$

$$= \Phi_A(s)\Phi_A^{-1}(\sigma(s)).$$

If $\Phi_A^{-1}(\sigma(s))$ is rd-continuous then

$$\Phi_A^{-1}(\sigma(s)) = \Phi_A^{-1}(s)$$

in which case

$$G(s^+, \sigma(s)) - G(s^-, \sigma(s)) = \Phi_A(s)\Phi_A^{-1}(s) = I.$$

(ii) The representation of $G(t, \sigma(s))$ by (4.6.4) shows that $G(t, \sigma(s))$ is a formal solution of (4.6.2), it fails to be a true solution because of discontinuity at $t = s. G(t, \sigma(s))$ satisfies homogeneous boundary conditions for,

$$\begin{aligned}
&MG(a, \sigma(s)) + NG(b, \sigma(s)) \\
&= -M\Phi_A(a)D^{-1}N\Phi_A(b, \sigma(s)) + N\Phi_A(b)D^{-1}M\Phi_A(a, \sigma(s)) \\
&= -M\Phi_A(a)D^{-1}N\Phi_A(b)\Phi_A^{-1}(\sigma(s)) \\
&\quad + N\Phi_A(b)D^{-1}M\Phi_A(a)\Phi_A^{-1}(\sigma(s)) \\
&= -M\Phi_A(a)D^{-1}[D - M\Phi_A(a)]\Phi_A^{-1}(\sigma(s)) \\
&\quad + N\Phi_A(b)D^{-1}M\Phi_A(a)\Phi_A^{-1}(\sigma(s)) \\
&= -M\Phi_A(a)\Phi_A^{-1}(\sigma(s)) + M\Phi_A(a)D^{-1}M\Phi_A(a)\Phi_A^{-1}(\sigma(s)) \\
&\quad + [D - M\Phi_A(a)]D^{-1}M\Phi_A(a)\Phi_A^{-1}(\sigma(s)) \\
&= -M\Phi_A(a, \sigma(s)) + M\Phi_A(a)D^{-1}M\Phi_A(a, \sigma(s)) \\
&\quad + M\Phi_A(a, \sigma(s)) - M\Phi_A(a)D^{-1}M\Phi_A(a, \sigma(s)) \\
&= 0.
\end{aligned}$$

Thus G is a formal solution of the boundary value problem.

(iii) Now to prove G is unique with properties (i) and (ii). Let $G_1(t,\sigma(s))$ be another Green's matrix with properties (i) and (ii).

Let $X(t,\sigma(s)) = G(t,\sigma(s)) - G_1(t,\sigma(s))$. At $t = s$ we have,

$$X(s^+,\sigma(s)) - X(s^-,\sigma(s))$$

$$= G(s^+,\sigma(s)) - G_1(s^+,\sigma(s)) - G(s^-,\sigma(s)) - G_1(s^-,\sigma(s))$$

$$= G(s^+,\sigma(s)) - G(s^-,\sigma(s)) - G_1(s^+,\sigma(s)) - G_1(s^-,\sigma(s))$$

$$= \Phi(s,\sigma(s)) - \Phi(s,\sigma(s))$$

$$= 0.$$

Thus $X(t,\sigma(s))$ has a removable discontinuity at $t = s$.

By regarding X appropriately we can make $X(t,\sigma(s))$ rd-continuous for all $t \in [a,b]$.

Now the boundary conditions are linear and X is a linear combination of G and G_1,

$$MX(a,\sigma(s)) + NX(b,\sigma(s))$$

$$= M[G(a,\sigma(s)) - G_1(a,\sigma(s))] + N[G(b,\sigma(s)) - G_1(b,\sigma(s))]$$

$$= [MG(a,\sigma(s)) + NG(b,\sigma(s))] - [MG_1(a,\sigma(s)) + NG_1(b,\sigma(s))]$$

$$= \Phi_A(s,\sigma(s)) - \Phi_A(s,\sigma(s))$$

$$= 0.$$

For each fixed s, $X(t, \sigma(s))$ is a solution of the homogeneous boundary conditions. Note that if $X(t, \sigma(s))$ is rd-continuous and possesses continuous first order derivatives at all points except at $t = s$, then $X(t, \sigma(s))$ is also a solution of the homogeneous boundary value problem.

4.7 Sturmian Theory

In this section, we present basic aspects of Sturmian theory for second order linear dynamic equations on measure chains. We shall first give some needed results for linear dynamic systems and then develop Sturmian theory. In addition to the results discussed in Section 2.5, we require the following results. Let $\mathbb{T}$ be an arbitrary measure chain and let us consider

$$x^{\Delta} = A(t)\, x + b(t), \tag{4.7.1}$$

$$x^{\Delta} = A(t)\, x\,, \tag{4.7.2}$$

where $A : \mathbb{T} \to B(X)$, $b : \mathbb{T} \to X$, X being a Banach space as in Section 2.5. We then define the generalized trace function $\mathbb{T}_h :$ $L(R^n) \to R$ given by

$$\mathbb{T}_h(A) = \lim_{k \to h} \frac{det(I+Ak)-1}{k} = \begin{cases} \frac{det(I+Ak)-1}{k} & if\ h > 0 \\ trace\, A, & if\ h = 0 \end{cases} \qquad (4.7.3)$$

The identity for $h = 0$ is a consequence of theorems on the Wronskian determinant in linear ODE theory. One can derive it by explicitly writing down the definition of the determinant via alternative sums.

Theorem 4.7.1: *Let X be finite-dimensional.*

(i) If x *is a solution of the matrix dynamic equation* (4.7.2), *then* $w(t) = det\ x(t)$ *is a solution of the scalar equation*

$$w^{\Delta} = \mathbb{T}_{\mu^*(t)}(A(t)) \cdot w\,. \qquad (4.7.4)$$

(ii) *If for all* $t < \tau$,

$$\mathbb{T}_{\mu^*(t)}(A(t)) \cdot \mu^*(t) + 1 \neq 0, \qquad (4.7.5)$$

then the IVP (4.7.1) *has exactly one solution on the whole measure chain* R.

Proof: (i) We show the identity (4.7.4) for $w(\,\cdot\,)$ for fixed $t \in \mathbb{T}$, one has to consider the two cases $\mu^*(t) > 0$ and $\mu^*(t) = 0$ separately. In the first case the assertion is trivial to obtain. For the second case we argue as follows. Let $t \in \mathbb{T}$ be right-dense, fixed and let l be the local

Lipschitz constant for $x(t)$. Since x is differentiable in t, for given $\epsilon > 0$ there is a neighborhood U_1 of t such that

$$|x(s) - x(t) - A(t)x(t)\mu(s,t)|_x \leq \tfrac{\epsilon}{2(l+1)} \cdot |\mu(s,t)|$$

or equivalently

$$|x(s) - [i\,d_x + A(t)\mu(s,t)] \cdot x(t)|_x \leq \tfrac{\epsilon}{2(l+1)} \cdot |\mu(s,t)|$$

for all $s \in U_1$. Applying det yields

$$|det\, x(s) - det\,[i\,d_x + A(t)\mu(s,t)] \cdot det\, x(t)| \leq \tfrac{\epsilon}{2} \cdot |\mu(s,t)|$$

for all $s \in U_2$, $U_2 \subseteq U_1$ an appropriate neighborhood of t. Now, according to (4.7.3) there is a third neighborhood $U_3 \subseteq U_2$ such that

$$|det\,[i\,d_x + A(t)\mu(s,t)] - 1 - \mathbb{T}_0(A(t))\mu(s,t)|$$

$$\leq \tfrac{\epsilon}{2(\,det|x(t)|+1)} \cdot |\mu(s,t)|.$$

Combining all these inequalities we obtain

$$|\,det\; x(s) - det\; x(t) - \mathbb{T}_0(A(t))det\, x(t)\mu(s,t)|$$

$$\leq |det\, x(s) - det\,[i\,d_x + A(t)\mu(s,t)] \cdot det\, x(t)|$$

$$+ |det\,[i\,d_x + A(t)\mu(s,t)] - 1 - \mathbb{T}_0(A(t))\mu(s,t)| \cdot |det\, x(t)|$$

$$\leq \epsilon \cdot |\mu(s,t)|.$$

This is the statement of the theorem.

(ii) Condition (4.7.5) is equivalent to $A(t)$ being regressive in Theorem 2.5.1 and hence the statement follows.

If we set $A \equiv 0$ in Theorem 2.5.1, we have a statement on the existence of antiderivatives.

Corollary 4.7.1: *If* $b : \mathbb{T} \to X$ *is rd-continuous, then there is an antiderivative*

$$B(t) = \int_{\tau}^{t} b(s)\,\Delta s \tag{4.7.6}$$

with the properties

$$B^{\Delta}(t) - b(t),\ t \in \mathbb{T},\ B(\tau) = 0, \tag{4.7.7}$$

by which it is uniquely determined.

For $\mathbb{T} = R$ this integral coincides with the ordinary integral for Banach-space valued functions, defined by Riemannian sums. For $\mathbb{T} = Z$ we have

$$\int_{\tau}^{t} b(s)\,\Delta s = sgn(t-\tau) \cdot \sum_{i=min\{t,\tau\}}^{max\{t,\tau\}-1} b(i).$$

We shall next present some theorems from calculus on measure chains which will be needed later. These theorems (except Theorem 4.7.3(ii)) are well-known for the time scales $\mathbb{T} = R$ or $\mathbb{T} = Z$.

Theorem 4.7.2: *Let* $\tau \in \mathbb{T}$ *and* $W : \mathbb{T}_\tau \to R$ *a rd-continuously differ-entiable strictly increasing positive function, with* $\lim_{t\to\infty} W(t) = \infty$ *(for example* $W(t) = \mu(t, \tau_1), \tau_1 > \tau$*). Then there is a positive constant* M *such that*

$$\int_\tau^\infty \frac{W^\Delta(s)\Delta s}{W^\alpha(s)} \begin{cases} = \infty, & if\, \alpha \le 1 \\ \le M \cdot sup\{t \in \mathbb{T}_\tau : W^\Delta(t)\mu^*(t) + 1\}, & if\, \alpha > 1. \end{cases} \tag{4.7.8}$$

Proof: $\alpha \le 1$: Assume to the contrary that $\int_\tau^\infty \frac{W^\Delta(s)\Delta s}{W^\alpha(s)} = N < \infty$. Then there is $\tau_1 > \tau$ with $\int_t^{\tau_1} \frac{W^\Delta(s)\Delta s}{W^\alpha(s)} > N - \frac{1}{2}$. Choose $\tau_2 > \tau_1$ such that $W(\tau_2) \ge min\{2W(\tau_1), 1\}$. Then

$$\int_\tau^{\tau_2} \frac{W^\Delta(s)\Delta s}{W^\alpha(s)} = \int_\tau^{\tau_1} \frac{W^\Delta(s)\Delta s}{W^\alpha(s)} + \int_{\tau_1}^{\tau_2} \frac{W^\Delta(s)\Delta s}{W^\alpha(s)}$$

$$> N - \tfrac{1}{2} + \int_{\tau_1}^{\tau_2} \frac{W^\Delta(s)\Delta s}{W^\alpha(s)} \ge N - \tfrac{1}{2} + \frac{W(\tau_2) - W(\tau_1)}{W^\alpha(\tau_2)}$$

$$\ge N - \tfrac{1}{2} + \tfrac{1}{2}\frac{W(\tau_2)}{W^\alpha(\tau_2)} \ge N,$$

noting $\frac{W(\tau_2)}{W^\alpha(\tau_2)} \ge 1$. This is a contradiction.

For the case $\alpha > 1$, we define $a_0 = b_0 = \tau$, $a_n = sup\{t \in \mathbb{T}_\tau, W(t) \le n\}$, $n \in N$, $b_n = inf\{t \in \mathbb{T}_\tau, W(t) \ge n\}$, $n \in N$.

These constants satisfy the following relations (for $n \in N$),

$$\lim_{n\to\infty} a_n = \lim_{n\to\infty} b_n = \infty,\ a_n \le b_n,$$

$$W(a_n) \le n \le W(b_n) \text{ if } a_n < b_n$$

then $\sigma(a_n) = b_n$.

From the last inequality we get for $n \in N$,

$$W(b_n) - W(a_n) \le W(\sigma(a_n)) - W(a_n) = W^{\Delta}(a_n)\mu^*(a_n)$$

$$\le sup\{t \in \mathbb{T}_\tau : W^{\Delta}(t)\mu^*(t)\}.$$

Now, we have for all n,

$$W(b_{n+1}) - W(b_n) = W(b_{n+1}) - W(a_{n+1}) + W(a_{n+1}) - W(b_n)$$

$$\le \sup_{t\in T} \{W^{\Delta}(t)\mu^*(t)\} + n + 1 - n$$

$$= \sup_{t\in T} \{W^{\Delta}(t)\mu^*(t) + 1\},$$

and further

$$\int_\tau^\infty \frac{W^{\Delta}(s)\Delta s}{W^{\alpha}(s)} = \sum_{n=0}^{\infty} \int_{b_n}^{b_{n+1}} \frac{W^{\Delta}(s)\Delta s}{W^{\alpha}(s)} \le \sum_{n=0}^{\infty} \int_{b_n}^{b_{n+1}} \frac{W^{\Delta}(s)\Delta s}{W^{\alpha}(b_n)}$$

$$\le \sum_{n=0}^{\infty} [W(b_{n+1}) - W(b_n)] \cdot \frac{1}{W^{\alpha}(b_n)}$$

$$\le sup\{t \in \mathbb{T}_\tau : W^{\Delta}(t)\mu^*(t) + 1\} \cdot \sum_{n=0}^{\infty} \frac{1}{W^{\alpha}(b_n)}$$

$$\le [\frac{1}{W^{\alpha}(b_n)} + \sum_{n=1}^{\infty} \frac{1}{n^{\alpha}}] \cdot sup\{t \in \mathbb{T}_\tau : W^{\Delta}(t)\mu^*(t) + 1\}.$$

Remark: The following example illustrates the case when $\alpha > 1$. Choose the measure chain $\{t_n\}_{n\in N_0}$, where the growth calibration is defined by

$$\mu(t_{n+1}, t_n) = \begin{cases} 1, & if \;\; n = 0 \\ \mu(t_n, t_0)^{\alpha}, & if \;\; n \geq 1. \end{cases}$$

Then we have

$$\int_{t_1}^{\infty} \frac{\Delta s}{\mu(t_n,t_0)^{\alpha}} = \sum_{n=1}^{\infty} \int_{t_n}^{t_{n+1}} \frac{\Delta s}{\mu(t_n,t_0)^{\alpha}} = \sum_{n=1}^{\infty} \frac{\mu(t_{n+1},t_n)}{\mu(t_n,t_0)^{\alpha}} = \infty.$$

So in the case $\alpha > 1$ of the above theorem one cannot expect boundedness of the infinite integral in general.

Theorem 4.7.3: *Let $g : \mathbb{T} \to R$ be a function*

(i) *We have*

$$\liminf_{t \to \infty} \; g(t) \leq \liminf_{t \to \infty} \;\; \mu^{-1}(t, \tau) \int_{\tau}^{t} g(s) \, \Delta s$$

$$\leq \limsup_{t \to \infty} \;\; \mu^{-1}(t, \tau) \int_{\tau}^{t} g(s) \, \Delta s \leq \limsup_{t \to \infty} \;\; g(t) \qquad (4.7.9)$$

In particular, $\lim_{t \to \infty} g(t)$ exists, then

$$\lim_{t \to \infty} \; \mu^{-1}(t, \tau) \int_{\tau}^{t} g(s) \, \Delta s = \lim_{t \to \infty} g(t). \qquad (4.7.10)$$

(ii) *If g is increasing, then for $\alpha \geq 0$,*

$$\limsup_{t \to \infty} \;\; \mu^{1-\alpha}(t, \tau) \, g(t)$$

$$\leq 3^{\alpha} \cdot \limsup_{t \to \infty} \;\; \mu^{-\alpha}(t, \tau) \int_{\tau}^{t} g(s) \, \Delta s \qquad (4.7.11)$$

Proof: (i) We omit this part of the proof. It is similar to the corresponding proof in the usual case for $\mathbb{T} = R$.

(ii) For each $t \geq \tau$ choose $\tilde{t} \geq t$ such that $\mu(t,\tau) \leq \mu(\tilde{t},\tau) \leq \mu(t,\tau) + \bar{\mu}$.

Then we have $\mu(\tilde{t},\tau) = \mu(\tilde{t},t) + \mu(t,\tau) \leq 2\mu(t,\tau) + \bar{\mu} \leq 3\mu(t,\tau)$, for all t with $\mu(t,\tau) \geq \bar{\mu} = sup\ \mu^*$. Now, observing that $\frac{1}{\mu^\alpha(t,\tau)} \leq 3^\alpha \frac{1}{\mu^\alpha(\tilde{t},\tau)}$ we estimate,

$$\mu^{1-\alpha}(t,\tau)\, g(t) = \frac{\mu(t,\tau)}{\mu^\alpha(t,\tau)} g(t) \leq 3^\alpha \frac{\mu(\tilde{t},t)}{\mu^\alpha(\tilde{t},\tau)} g(t)$$

$$= \frac{3^\alpha}{\mu^\alpha(\tilde{t},\tau)} \int_t^{\tilde{t}} g(s)\,\Delta s \leq \frac{3^\alpha}{\mu^\alpha(\tilde{t},\tau)} \int_\tau^{\tilde{t}} g(s)\,\Delta s$$

Now, taking $\limsup_{t\to\infty}$ yields the claimed statement.

The last lemma to be presented here deals with a technical feature concerning left sided derivatives in ldrs-points. It is only needed once in the proof of the Sturmian Comparison Theorem 4.7.5. It is interesting in so far as it reveals one of the technical problems which appear when one deals with general measure chains with non-constant graininess.

Let $f: \to X$ be a function and $t \in \mathbb{T}$ be left-dense. If the restriction $f|_{(-\infty,t]}$ of f to $(-\infty, t]$ is differentiable at t, then

$$f^{\Delta_l}(t) = (f|_{(-\infty,t]})^\Delta(t) \tag{4.7.12}$$

is called the *left-sided derivative of* f at t. We will need the following technical lemma.

Lemma 4.7.1 (Left sided derivative): *Let* $f : \to X$ *be rd-continuously differentiable and* $t \in \mathbb{T}$ *be left-dense.*

(i) *Then* f *has a left-sided derivative at* t,

$$f^{\Delta_l}(t) = \lim_{s \to t} f^{\Delta}(s) \tag{4.7.13}$$

(ii) *If* f *is continuously differentiable, we have*

$$f^{\Delta_l}(t) = f^{\Delta}(t)\,. \tag{4.7.14}$$

Remark: For the function $f : (-\infty, 0\ ] \ \cup N \to R$ given by $f(t) = |t|$ we get

$$f^{\Delta_l}(t) = \lim_{s \to 0} f^{\Delta}(s) = -1 \neq 1 = f^{\Delta}(0).$$

The left-sided derivative and the derivative do not coincide at the ldrs-point 0.

Proof: The function $g : (-\infty, t\,] \to X$ defined by

$$g(r) = \begin{cases} f^{\Delta}(r), & r \neq t \\ \lim\limits_{s \to ()} f^{\Delta}(s), & r = t \end{cases},$$

is rd-continuous (on $(-\infty, t]$) and continuous at t, because f^{Δ} is rd-continuous. g has a differentiable antiderivative $G : (-\infty, t] \to X$, which coincides with f on $(-\infty, t)$. Because f and G are continuous and t is left-dense, we even have $f(s) = G(s)$ for $s \in (-\infty, t]$. Therefore $f|_{(-\infty,t]} = G$ is differentiable at t and

$$f^{\Delta_l}(t) := (f|_{(-\infty,t]})^{\Delta}(t) = G^{\Delta}(t) = g(t) = \lim_{s \to t} g(s) = \lim_{s \to t} f^{\Delta}(s).$$

We shall now discuss the Sturm-Liouville equation.

From now on limit operators like lim, limsup.....always mean $\lim\limits_{t \to \infty}$, $\limsup\limits_{t \to \infty}$ .... We omit the sub line $t \to \infty$. For the rest of this section let two functions $r \in C^1_{rd}(\mathbb{T}, R^+)$, $q \in C_{rd}(\mathbb{T}, R)$ be fixed. The Sturm-Liouville operator for functions $x \in C^2_{rd}(\mathbb{T}, R)$ is defined by

$$L[x](t) := (r.x^{\Delta})^{\Delta}(t) + q(t).x^{\sigma}(t) \tag{4.7.15}$$

In this section we will study the homogeneous Sturm-Liouville equation

$$L[x](t) = 0 \tag{4.7.16}$$

For two functions $x, y \in C^2_{rd}(\mathbb{T}, R)$ the system $L[y](t) = 0, L[x](t) = 0$ is equivalent to the first order matrix equation

$$\begin{pmatrix} y & x \\ r\,y^{\Delta} & rx^{\Delta} \end{pmatrix}^{\Delta} = \begin{pmatrix} 0 & \frac{1}{r(t)} \\ -\,q(t) & \frac{-q(t)\mu^{*}(t)}{r(t)} \end{pmatrix} \cdot \begin{pmatrix} y & x \\ r\,y^{\Delta} & rx^{\Delta} \end{pmatrix}$$

$$= A(t)\begin{pmatrix} y & x \\ r\,y^{\Delta} & rx^{\Delta} \end{pmatrix}. \qquad (4.7.17)$$

(Observe that $x^{\sigma}(t) = x(t) + x^{\Delta}(t)\mu^{*}(t)$). By Theorem 4.7.1 the determinant of the Wronskian

$$w(t) : \;= W[x,y](t) = r(t)[x^{\Delta}y - y^{\Delta}x](t), \qquad (4.7.18)$$

of the solution matrix is a solution to the scalar equation

$$w^{\Delta} = \mathbb{T}_{\mu^{*}(t)}(A(t)) \cdot w = 0,$$

and therefore constant. By (ii) of this theorem the Sturm-Liouville equation (4.7.16) and the inhomogeneous equation

$$L[\,x\,](t) = h(t) \qquad (4.7.19)$$

have unique solutions on the whole measure chain $\mathbb{T}$, in particular in backward time direction.

The above considerations reveal a close connection between the two operators L and W. There are further connections which are established in the following identities. They are simple consequences of the product rule.

Lemma 4.7.2: *Let* $x, y \in C^2_{rd}(\mathbb{T}, R)$. *We then have*

(i) <u>*Lagrange identity*</u>

$$W[x,y]^{\Delta}(t) = (L[x]\cdot y^{\sigma} - L[y]\cdot x^{\sigma})(t) \tag{4.7.20}$$

(ii) <u>*Green's formula*</u>

$$W[x,y](t) - W[x,y](\tau)$$

$$= \int_{\tau}^{t}(L[x](s)\cdot y^{\sigma}(s) - L[y](s)\cdot x^{\sigma}(s))\Delta s. \tag{4.7.21}$$

(iii) <u>*Boundary Value Problem*</u>

If the functionals $B_a^+, B_a^- : C^1_{rd}(\mathbb{T}, R) \to R,\ a \in \mathbb{T}$ *are defined by*

$$B_a^+[x] = \alpha_1 x(a) + \alpha_2 r(a)\, x^{\Delta}(a), \tag{4.7.22}$$

$$B_a^-[x] = \alpha_2 x(a) - \alpha_1 r(a)\, x^{\Delta}(a), \tag{4.7.23}$$

then

$$(\alpha_1^2 + \alpha_2^2)\cdot W[x,y](a) = B_a^+[x]\cdot B_a^-[y] - B_a^+[y]\cdot B_a^-[x]. \tag{4.7.24}$$

Proof: The Green's formula is the integrated form of the Lagrange identity. Equation (4.7.24) can be proven by straightforward calculations. Because of its importance we explicitly show the Lagrange identity.

$$W[x,y]^{\Delta}(t) = [r\,x^{\Delta}\cdot y]^{\Delta}(t) - [r\,y^{\Delta}\cdot x]^{\Delta}(t)$$

$$= [(r\,x^{\Delta})^{\Delta}(t)\cdot y^{\sigma}(t) + (r\,x^{\Delta}y^{\Delta})(t)]$$

$$- [(r\,y^{\Delta})^{\Delta}(t)\cdot x^{\sigma}(t) + (r\,y^{\Delta}x^{\Delta})(t)]$$

$$= [(r\,x^{\Delta})^{\Delta}(t)\cdot y^{\sigma}(t) + (q\,x^{\sigma}y^{\sigma})(t)]$$

$$- [(r\,y^{\Delta})^{\Delta}(t)\cdot x^{\sigma}(t) + (q\,y^{\sigma}x^{\sigma})(t)]$$

$$= L[x](t)\cdot y^{\sigma}(t) - L[y](t)\cdot x^{\sigma}(t).$$

Remarks: (a) From (ii) and (iii) one sees that two functions x and y, both subject to the boundary conditions $B_a^+[x] = 0 = B_b^+[x]$, $a, b \in \mathbb{T}$ fulfill the identity

$$\int_a^b (L[x](s)\cdot y^{\sigma}(s) - L[y](s)\cdot x^{\sigma}(s))\Delta(s) = 0.$$

(b) From the Lagrange identity we again see that two $C^2_{rd} - solutions$ of (4.7.16) satisfy $W[x,y] = const.$

In the following we will consider the relation $W[x,y] = const.$ or the even weaker relation $sgn\,W[x,y] = const.$ instead of the homogeneous Sturm-Liouville equation (4.7.16). This will be done for a special class of functions which we define now:

A $C^0_{rd} - function\, x : \mathbb{T} \to R$ is called $nonoscillatory$(with $initial\, time\ \tau \in \mathbb{T}$), if $x^{\sigma}(t)x(t) > 0$ for $t \in \mathbb{T}$. Each $\tau_1 \geq \tau$ is also an initial time. A function which is not nonoscillatory, is called $oscillatory$. From the intermediate value theorem it follows that a nonoscillatory function does not change its sign for times greater than the initial time. The importance of nonoscillatory functions comes from the fact that the quotient rule relates the quotient y/x of two nonoscillatory functions x and y to the Wronskian operator $W[x, y]$.

Lemma 4.7.3: *Let x be nonoscillatory with initial time τ and $y \in C^1_{rd}(\mathbb{T}, R)$.*

(*i*) *There is an equivalence*

$$y(t) = \left\{\frac{y(\tau)}{x(\tau)} + \int_\tau^t \frac{f(s)\Delta s}{r(s)x^\sigma(s)x(s)}\right\}x(t),\ t \in \mathbb{T}_\tau$$

$$\Leftrightarrow f(s) = W[y, x](s),\quad s \in \mathbb{T}_\tau \tag{4.7.25}$$

Therefore x, y are linearly dependent, iff $W[x, y] \equiv 0$.

(ii) *If $W[y, x] \leq 0$, $\tau \in T$ or $W[y, x] \geq 0$, $\tau \in T$, then y is eventually identically zero or also nonoscillatory.*

Proof: (i) is a simple consequence of the quotient rule. If the sign condition in (ii) is fulfilled, then the function in the braces in (4.7.25) is nondecreasing or nonincreasing and therefore eventually zero, positive or negative. So, if y is not eventually zero, it is, together with x, nonoscillatory.

A C^0_{rd} $-$ *function* $x : \mathbb{T} \to R$ is called *dominant* or *reces-sive* (with initial time τ), if it is nonoscillatory (with initial time τ) and

$$\int_\tau^\infty \frac{\Delta s}{r(s)x^\sigma(s)x(s)} < \infty \quad or \quad \int_\tau^\infty \frac{\Delta s}{r(s)x^\sigma(s)x(s)} = \infty \ ,$$

respectively.

Lemma 4.7.4: *Let $x,\ y \in C^1_{rd}(\mathbb{T}, R)$ be given.*

(i) *If x is recessive and $W[y,x]$ is bounded away from zero, then y is dominant. Further, x and y satisfy* $\lim \frac{x}{y}(t) = 0$ *and* $\frac{r\,x^\Delta}{x}(t) < \frac{r\,y^\Delta}{y}(t),\ t \in T.$

(ii) *If $W[y,x]$ is bounded and* $\lim \frac{x}{y}(t) = 0$, *then y is recessive or trivial.*

Proof: If x and y are nonoscillatory with initial time τ, then by (4.7.25) for $t \geq \tau$ we have the equation

$$\frac{y(\tau)}{x(\tau)} + \int_\tau^t \frac{W[x,y](s)\Delta s}{r(s)y^\sigma(s)y(s)} = \frac{x(t)}{y(t)} = \left(\frac{y(t)}{x(t)}\right)^{-1}$$

$$= \left[\frac{y(\tau)}{x(\tau)} + \int_\tau^t \frac{W[y,x](s)\Delta s}{r(s)x^\sigma(s)x(s)}\right]^{-1}. \qquad (4.7.26)$$

We prove (i). From Lemma 4.7.3 (ii) we know that y is nonoscillatory. In equation (4.7.26) the right hand side goes to zero for $t \to \infty$. Now, considering the left part of (4.7.26) obtain that y is dominant and satisfies the limit property. Moreover, we get for $t \geq \tau$, $sgn\, W[x,y](t) = -\, sgn \frac{x(\tau)}{y(\tau)} = -\, sgn \frac{x(t)}{y(t)} = -\, sgn\, x(t)y(t)$. Therefore, it follows that $\frac{r(t)x^{\Delta}(t)}{x(t)} - \frac{r(t)y^{\Delta}(t)}{y(t)} = \frac{W[x,y](t)}{x(t)y(t)} < 0$.

(ii) is an immediate consequence of the first equation in (4.7.26).

The proof is complete.

Theorem 4.7.4 (Nonoscillatory solutions of the Sturm-Liouville equation):

(i) *If* x *is a nonoscillatory solution of* (4.7.16), *then the following statements for a function* $y \in C^2_{rd}(\mathbb{T}, R)$ *are equivalent*

y *is a nonoscillatory solution of* (4.7.16)

y *is a nontrivial solution of* (4.7.16)

$W[x,y](t) \equiv const \neq 0$

(ii) <u>*Existence*</u>: *If* $x : \mathbb{T}_\tau \to R$ *is a recessive solution of* (4.7.16), *then* $y : \mathbb{T}_\tau \to R$, *defined by*

$$y(t) = x(t) \cdot \int_\tau^t \frac{\Delta s}{r(s)x^\sigma(s)x(s)} \tag{4.7.27}$$

is a dominant solution. If $x : \mathbb{T}_\tau \to R$ *is a dominant solution of* (4.7.16), *then* $y : \mathbb{T}_\tau \to R$, *defined by*

$$y(t) = x(t) \cdot \int_\infty^t \frac{\Delta s}{r(s)x^\sigma(s)x(s)} \tag{4.7.28}$$

is a recessive solution. In particular, there are always recessive and dominant solutions, if there are nonoscillatory ones.

(iii) *Characterization of recessive solutions: The following statements for a solution* $y : \mathbb{T}_\tau \to R$ *are equivalent:*

(A) y *is recessive.*

(B) $\frac{y}{x} = const.$ *for all recessive solutions* x.

(C) $\lim \frac{y}{x}(t) = 0$ *for all dominant solutions* x.

From $(A) \Leftrightarrow (B)$ *one can see that the space of all recessive solutions is* 1*-dimensional.*

(iv) *Characterization of dominant solutions: The following statements for a solution* $y : \mathbb{T}_\tau \to R$ *are equivalent:*

(D) y *is dominant.*

(E) $\lim \frac{y}{x}(t) \neq 0$ *for all dominant solutions* x.

(F) $\lim \frac{x}{y}(t) = 0$ *for all recessive solutions* x.

Proof: The equivalences in (i) are immediate consequences of the Lagrange identity (Lemma 4.7.2 (i)).

(ii) First part: From Lemma 4.7.3(i) we know that $W[y,x] \equiv 1$. By (i) y is a solution, and by Lemma 4.7.4 (i) y is dominant. For the second part, we write y as

$$y(t) = x(t) \cdot \int_{\infty}^{t} \frac{\Delta s}{r(s)x^{\sigma}(s)x(s)} = \left\{ \frac{y(\tau)}{x(\tau)} + \int_{\tau}^{t} \frac{\Delta s}{r(s)x^{\sigma}(s)x(s)} \right\} x(t).$$

So by Lemma 4.7.3 we again have $W[y,x] \equiv 1$ and y is a solution. From Lemma 4.7.4 (ii) we see that y is recessive.

(iii) $(A) \Rightarrow (B)$: Assume to the contrary that x, y are recessive solutions with $\frac{y}{x} \neq const.$ then we have $W[y,x] = const \neq 0$. From Lemma 4.7.3 (i) we get

$$\frac{y(t)}{x(t)} = \left[\frac{y(\tau)}{x(\tau)} + W[y,x] \cdot \int_{\tau}^{t} \frac{\Delta s}{r(s)x^{\sigma}(s)x(s)} \right] \to \pm\infty,$$

$$\frac{x(t)}{y(t)} = \left[\frac{x(\tau)}{y(\tau)} + W[x,y] \cdot \int_{\tau}^{t} \frac{\Delta s}{r(s)y^{\sigma}(s)y(s)} \right] \to \pm\infty,$$

This is impossible.

$(B) \Rightarrow (A)$ is trivial.

Now we consider the identity

$$\frac{x(t)}{y(t)} - \frac{x(\tau)}{y(\tau)} = W[x,y] \cdot \int_{\tau}^{t} \frac{\Delta s}{r(s)y^{\sigma}(s)y(s)}.$$

$(C) \Rightarrow (A)$: The left hand side goes to $\pm\infty$ for all dominant solutions x. So y must be recessive.

$(A) \Rightarrow (C)$: If y is recessive and x dominant, then $W[x,y] \neq 0$. Therefore $\lim \frac{x}{y}(t) = \pm\infty$. Thus $\lim \frac{y}{x}(t) = 0$.

(iv) $(D) \Rightarrow (F)$ is identical to $(A) \Rightarrow (C)$.

$(F) \Rightarrow (D)$: If y would be recessive, then by $(A) \Leftrightarrow (B)$ and (F) : $\frac{x}{y} = const. = 0$. So x is trivial. We have a contradiction.

$(D) \Rightarrow (E)$: Because of the equation $\frac{x(t)}{y(t)} = \frac{x(\tau)}{y(\tau)} + W[x,y] \cdot \int_\tau^t \frac{\Delta s}{r(s)y^\sigma(s)y(s)}$, if x is dominant then $\lim \frac{x}{y}(t)$ exists and is finite. By symmetry $\lim \frac{y}{x}(t)$ also exists and is finite. Therefore $\lim \frac{y}{x}(t) \neq 0$ exists.

$(E) \Rightarrow (D)$: Assume y is recessive. Then by (ii), $(A) \Rightarrow (C)$ we have $\lim \frac{y}{x}(t) = 0$. This is again a contradiction.

From (i) in the preceding theorem we see that either all nontrivial solutions are nonoscillatory or none is. This motivates the following definition: Equation (4.7.16) is called *nonoscillatory*, if there is a nonoscillatory solution. Otherwise (4.7.16) is called *oscillatory*. A much more precise statement about nodes of different solutions of (4.7.16) is contained in the following central theorem and its corollaries.

Theorem 4.7.5 (Sturmian Comparison Theorem): *Let $a < b \in \mathbb{T}$ and let $x, y \in C_{rd}^2(\mathbb{T}, R)$ fulfill the following conditions*

(A) $\frac{L[x](t)}{x^\sigma(t)} \leq \frac{L[y](t)}{y^\sigma(t)}$ *for all $t \in [a,b)$ with $x^\sigma(t) \cdot y^\sigma(t) \neq 0$.*

(B) *y has nodes at a and at b.*

(C) *y has no nodes in (a,b).*

(D) $y(b) \neq 0$, *if* $(a, b) = \emptyset$.

Then x *has a node in* $[a, b]$.

Remark: If $y(b) = 0$ for a left-scattered $b \in \mathbb{T}$, then $y(\rho(b)) \cdot y(b) = 0$ (ρ is the backward jump operator). The zero at b causes two nodes at $\rho(b)$ and at b. By assumption (D) we exclude the case that the nodes at a and at b are of this kind. Without condition (D) the theorem would become false. As a counter-example take, $\mathbb{T} = Z$, $L[x] = x^{\Delta\Delta}$, $a = -1$, $b = 0$, $y(t) = t$, $x(t) = 1$.

If y is oscillatory, there will always (for $t \to \infty$) be pairs of nodes, which fulfill conditions $(A) - (D)$.

Proof: (a) Assume that x has no nodes in $[a, b]$. Then without loss of generality

$$x(t) > 0 \quad \text{for } t \in [a, \sigma(b)]. \tag{4.7.29}$$

From (C) and (D) we see that (gain w.l.o.g)

$$y(t) > 0 \text{ for } t \in J = \begin{cases} (a, b), & if\ b\ is\ left-dense, \\ (a, b), & if\ b\ is\ left-scattered. \end{cases} \tag{4.7.30}$$

(b) We show that the statement $I(t)$

$$I(t) : L[x](t) \cdot y^{\sigma}(t) \le L[y](t) \cdot x^{\sigma}(t) \qquad (4.7.31)$$

holds for all $t \in [a, b)$ with $\sigma(t) \in J$. If b is left-dense, then $\sigma(t) \in (a, b) = J$ for all $t \in (a, b)$. If b is left-scattered, then $\sigma(t) \in (a, b] = J$ for all $t \in (a, b)$. This yields $I(t)$ for $t \in (a, b)$. If a is right-scattered, then $\sigma(a) \in J$, we get $I(a)$. If a is right-dense, then we get $I(a)$, because the functions in the inequality $I(t)$ are continuous at a.

(c) We define the function $W : \mathbb{T} \to R$ by $W(t) = W[y, x](t) = r(t) \cdot [y^{\Delta} \cdot x - x^{\Delta} y](t)$. W is rd-continuously differentiable, because x and y as solutions of (4.7.16) are twice rd-continuously differentiable. For right-scattered t we get,

$$W(t) = \frac{r(t)}{\mu^{*}(t)} \cdot [x(t)y^{\sigma}(t) - y(t)x^{\sigma}(t)].$$

(d) W is monotone increasing in $[a, b]$, because for $t \in [a, b)$ we have Lagrange's identity and (b) : $W^{\Delta}(t) - (L[y]x^{\sigma} - L[x]\, y^{\sigma})(t) \ge 0$.

(e) We have $W(a) \ge 0$. Consider the two cases

Case 1: a right-dense, $W(a) = r(a) \cdot [x(a)y^{\Delta}(a) - y(a)x^{\Delta}(a)] \ge 0$, since $x(a) \ge 0$, $y^{\Delta}(a) \ge 0$ and $y(a)$=0.

Case 2: a right-scattered, $W(a) = \frac{r(a)}{\mu^*(a)} \cdot [x(a)y^\sigma(a) - y(a)x^\sigma(a)]$ ≥ 0, because $x(a) \geq 0$, $y^\sigma(a) \geq 0$, $y(a) \leq 0$ and $x^\sigma(a) \geq 0$.

The inequalities for y follow from (4.7.30) and the fact that y has a node at a.

(f) We have $W(b) \leq 0$. From (4.7.30) we conclude that $y(b) \geq 0$. Here we consider the two main cases

Case 1: $y(b) = 0$.

Case 1a: b right-dense. Then $W(t) = \frac{r(b)}{\mu^*(b)} \cdot [x(b)y^\sigma(b) - y(b)x^\sigma(b)]$ $= 0$, since $y^\sigma(b) = 0$, and $y(b) = 0$.

Case 1b: b right-scattered and left-dense. By Lemma 4.7.1 (y^Δ is continuous) we have

$$\frac{y^\sigma(b)}{\mu^*(b)} = \frac{y^\sigma(b) - y(b)}{\mu^*(b)} = y^\Delta(b) = y^{\Delta_l}(b) : = \left(y\big|_{(-\infty,b]}\right)^\Delta(b) \leq 0,$$

by (4.7.30) and therefore $W(b) = \frac{r(b)}{\mu^*(b)} \cdot x(b)y^\sigma(b) \leq 0$.

Case 1c: b right-scattered and left-scattered. Here we have $y(\rho(b)) \cdot y(b) = 0$ and $\rho(b)$ would be a node which is in (a, b) (cf. conditions $(C), (D)$). This case is impossible.

Case 2: $y(b) > 0$. b is a node and thus b must be right-scattered with $y^\sigma(b) \leq 0$. Hence we get, noting that $x(b) \geq 0$, $y^\sigma(b) \leq 0$, $y(b) > 0, x^\sigma(b) > 0$,

$$W(b) = \frac{r(b)}{\mu^*(b)} \cdot [x(b)y^\sigma(b) - y(b)x^\sigma(b)] \leq 0.$$

(g) By (d), (e), (f) W is vanishing in $[a, b]$. Thus, for $t \in [a, b]$ we get,

$$\left(\frac{y}{x}\right)^\Delta(t) = \frac{y^\Delta(t)x(t) - x^\Delta(t)y(t)}{x^\sigma(t)x(t)} = \frac{W[y,x](t)}{r(t)x^\sigma(t)x(t)} = 0.$$

So $\frac{y}{x}$ is a constant on $[a, \sigma(b)]$ and by condition (B) x has nodes at a and at b. This is a contradiction to the assumption at the beginning of the proof.

Corollary 4.7.2: *Let* $a, b \in \mathbb{T}$. *Assume that there is a solution* $y : \mathbb{T} \to R$ *of* (4.7.16) *which fulfills conditions* $(A) - (D)$ *of Theorem* 4.7.5. *Then every solution* x *of the homogeneous Sturm-Liouville equation* (4.7.16) *has a node in* $[a, b]$.

Corollary 4.7.3 (Nonoscillation Criterion): *If there is a non-vanishing function* $x : \mathbb{T}_\tau \to R$ *with*

$$\frac{L[x](t)}{x^\sigma(t)} \leq 0, \quad t \in \mathbb{T}_\tau,$$

then equation (4.7.16) *is nonoscillatory.*

Proof: For an oscillatory solution y one can find a pair of nodes (a, b) such that all conditions of Theorem 4.7.5 are fulfilled. This is contradiction.

Let us next consider the Riccati equation.

Let again $r \in C^1_{rd}(\mathbb{T}, R)$ and $q \in C^2_{rd}(\mathbb{T}, R)$ be given. For $\tau \in \mathbb{T}$ and $C^1_{rd} - functions\, u : \mathbb{T}_\tau \to R$ with $r(t) + u(t)\mu^*(t) > 0$ we define the operator

$$P[u] : \begin{cases} \mathbb{T}_\tau \to R_0^+ \\ t \ \mapsto \ \frac{u^2(t)}{r(t)+u(t)\mu^*(t)} \end{cases} \tag{4.7.32}$$

and the Riccati operator

$$R[u] : \begin{cases} \mathbb{T}_\tau \to R \\ t \ \mapsto u^\Delta(t) + P[u](t) + q(t). \end{cases} \tag{4.7.33}$$

We shall study the Riccati equation

$$R[u] = u^\Delta + P[u](t)78 + q(t)$$

$$= 0.u^{\Delta} + \frac{u^2}{r(t)+u\,\mu^*(t)} + q(t) = 0 \tag{4.7.34}$$

The following statements for a given function u will play an important role,

(U_1) There is $\tau \in \mathbb{T}$ and a function $u : \mathbb{T}_\tau \to R$ with $r(t) + u(t)\mu^*(t) > 0$ and $R[u](t) = 0,\ t \in \mathbb{T}_\tau$.

(U_2) There is $\tau \in \mathbb{T}$ and a function $u : \mathbb{T}_\tau \to R$ with $r(t) + u(t)\mu^*(t) > 0$ and $R[u](t) \leq 0,\ t \in \mathbb{T}_\tau$.

(U_3) There is $\tau \in \mathbb{T}$, $c \in R$ and a function $u : \mathbb{T}_\tau \to R$ with $r(t) + u(t)\mu^*(t) > 0$ and $u(t) = c - \int_\tau^t (P[u](s)\, q(s))\, \Delta s$.

(U_4) There is $\tau \in \mathbb{T}$, $c \in R$ and a function $u : \mathbb{T}_\tau \to R$ with $r(t) + u(t)\mu^*(t) > 0$ and $u(t) \geq c - \int_\tau^t (P[u](s)\, q(s))\, \Delta s \geq 0$ or $u(t) \leq c - \int_\tau^t (P[u](s)\, q(s))\, \Delta s \leq 0$ for $t \in T_\tau$.

The relation between the Riccati equation (4.7.34) and the Sturm-Liouville equation (4.7.16) given in the following theorem, where (O_1) is the statement: Equation (4.7.16) is nonoscillatory.

Theorem 4.7.5(a): *The following implications are valid*

$$(O_1) \Leftrightarrow (U_1) \Leftrightarrow (U_2) \Leftrightarrow (U_3) \Leftarrow (U_4).$$

Remarks: (a) If $\int_{\tau}^{t} q(s)\,\Delta s \to \infty$, we can easily see that $(U_2) \Rightarrow (U_4)$ and so all statements are equivalent.

(b) We emphasize that by equivalence $(O_1) \Leftrightarrow (U_1)$ equation (4.7.16) is nonoscillatory, iff the Riccati equation has a 'solution' which is global and satisfies the inequality

$$r(t) + u(t)\mu^*(t) > 0 \tag{4.7.35}$$

For the two measure chains $\mathbb{T} = R$ and $\mathbb{T} = Z$ this condition (U_1) has two different meanings: For $\mathbb{T} = R$ the inequality is redundant. Here the fact that the nonlinear Riccati equation has a global solution ensures the nonoscillatory behavior of (4.7.16). For $\mathbb{T} = Z$ it is clear that (4.7.34) has always a global solution. Here the inequality (4.7.35) is necessary for (4.7.16) to be nonoscillatory. We will see that for time scales with nonvanishing graininess, this inequality is crucial for many nonoscillation criteria, which cannot be established in the continuous time case.

Proof: (a) We first show for $\tau \in \mathbb{T}$, if the functions

$$u : \mathbb{T}_\tau \to R\,, \quad r(t) + u(t)\mu^*(t) > 0\,, \tag{4.7.36}$$

$$y : \mathbb{T}_\tau \to R\,,\ y(t) > 0, \tag{4.7.37}$$

are related by y which is a solution of the IVP

$$x^{\Delta} = \frac{u(t)}{r(t)}x, \; x(\tau) = 1, \tag{4.7.38}$$

and

$$u(t) = \frac{r(t)y^{\Delta}(t)}{y(t)}, \tag{4.7.39}$$

then the identity

$$\frac{L[y](t)}{y^{\sigma}(t)} = R[u](t), \;\; t \in \mathbb{T}_{\tau} \tag{4.7.40}$$

holds. This is an easy consequence of the product rule,

$$L[y](t) = (r \cdot y^{\Delta})^{\Delta}(t) + q(t)y^{\sigma}(t) = (u \cdot y^{\Delta})^{\Delta}(t) + q(t)y^{\sigma}(t)$$

$$= y^{\sigma}(t) \cdot u^{\Delta}(t) + y^{\Delta}(t) \cdot u(t) + q(t)y^{\sigma}(t)$$

$$= y^{\sigma}(t) \cdot [\, u^{\Delta}(t) + \frac{u^2(t)}{r(t)+u(t)\,\mu^*(t)} + q(t)],$$

because $y^{\Delta}(t) = \frac{u(t)}{r(t)}y(t) = y^{\sigma}(t) \cdot \frac{u(t)}{r(t)} \cdot \frac{y(t)}{y^{\sigma}(t)} = y^{\sigma}(t) \cdot \frac{u(t)}{r(t)} \cdot \frac{1}{1+\frac{u(t)}{r(t)}\mu^*(t)}$.

(b) The implications $(U_3) \Leftrightarrow (U_1) \Rightarrow (U_2)$are trivial. From the remarks in (a) it is easily seen that (U_1) is equivalent to (O_1). Equation (4.7.16) is nonoscillatory. If (U_2) holds, we define by (4.7.38) and then conclude by (4.7.40) that $\frac{L[y](t)}{y^{\sigma}(t)} \leq 0, \; t \in \mathbb{T}_{\tau}$.

Now Corollary 4.7.3. assures that equation (4.7.16) is nonoscillatory.

(c) It remains to be shown that (U_4) implies the other statements of the theorem. For $v(t) :\ = c - \int_{\tau}^{t} (P[u](s)\, q(s))\, \Delta s$ we have

$$r(t) + v(t)\, \mu^*(t) \geq r(t) + min\{0, u(t)\}\mu^*(t) > 0$$

and

$$P[v](t) = \frac{v(t)}{\frac{r(t)}{v(t)} + \mu^*(t)} \leq \frac{u(t)}{\frac{r(t)}{u(t)} + \mu^*(t)} = P[u](t).$$

Thus we arrive at $R[v](t) = \ v^{\Delta}(t) + P[v](t) + q(t) \leq 0$, and therefore (U_3) holds.

We are next going to derive oscillation criteria via the equivalences listed in the above theorem. In these oscillation criteria there will appear various conditions concerning the two given functions q and r and certain properties of solutions of the Riccati equation. In order to simplify the statements we introduce several conditions.

The function $\delta^* : T \to \{0, 1\}$ is given by

$$\delta^*(t) = sgn\, \mu^*(t) = \begin{cases} 0, & if\ \sigma(t) = t, \\ 1, & if\ \sigma(t) > t. \end{cases}$$

List of Conditions for r :

(R_1) r is bounded.

(R_2) $lim\, sup\, \mu^{-1}(t, \tau) \int_{\tau}^{t} r(s) \quad \Delta s < \infty$ and $r \cdot \delta^*$ is bounded.

(R_3) $lim\, sup\, \mu^{-1}(t, \tau) \int_{\tau}^{t} r(s) \quad \Delta s < \infty$.

(R_4) $lim\,sup\,\mu^{-2}(t,\tau)\int_{\tau}^{t} r(s) \quad \Delta s < 0.$

(S_1) $\delta^* \equiv 1$ and $\frac{r}{\mu^*}$ bounded.

(S_2) $\delta^* \equiv 1$ and $lim\,sup\,\mu^{-1}(t,\tau)\int_{\tau}^{t} \frac{r(s)}{\mu^*(s)}\Delta s < \infty$ and r bounded.

(S_3) $\delta^* \equiv 1$ and $lim\,sup\,\mu^{-1}(t,\tau)\int_{\tau}^{t} \frac{r(s)}{\mu^*(s)}\Delta s < \infty.$

(S_4) $\delta^* \equiv 1$ and $lim\,sup\,\mu^{-\frac{3}{2}}(t,\tau)\int_{\tau}^{t} \frac{r(s)}{\mu^*(s)}\Delta s < \infty.$

(S_5) $\delta^* \equiv 1$

List of Conditions for *q* :

For fixed $T \in \mathbb{T}$ we define the integral $Q(t) : = \int_T^t q(s)\,\Delta s.$

$(Q_1) lim\,Q(t)$ exists and is finite.

$(Q_2) lim\,\mu^{-1}(t,\tau)\int_{\tau}^{t} Q(s)\Delta s$ exists and is finite.

$(Q_3) lim\,inf\,\mu^{-1}(t,\tau)\int_{\tau}^{t} r(s)\Delta s > -\infty.$

$(Q_4) lim\,sup\,\mu^{-1}(t,\tau)\int_{\tau}^{t} r(s)\Delta s > -\infty.$

$(Q_5) lim\,sup\,Q(t) > -\infty.$

$(Q_6) lim\,Q(t) = \infty.$

$(Q_7) inf\,Q(t)\delta^*(t) > -\infty.$

Because of

$$\mu^{-1}(t,\tau)\int_T^t \int_T^s - sTq(\overline{s}\ )\Delta\overline{s}\ \Delta s$$

$$= \frac{\mu(t,T_2)}{\mu(t,T)} \cdot \mu^{-1}(t,T_2) \cdot \int_{T_2}^{t} \int_{T_2}^{s} q(\overline{s}\,)\Delta\overline{s}\;\Delta s$$

$$- \mu^{-1}(t,T) \cdot \int_{T_2}^{t} \int_{T_2}^{s} q(\overline{s}\,)\Delta\overline{s}\;\Delta s - \int_{T_2}^{T} q(\overline{s}\,)\Delta s,$$

we can replace T by any arbitrary fixed time T_2 in the above conditions. These conditions are independent of T. In case $\mathbb{T}$ is a group one usually sets $T = 0$.

List of Conditions for a Given Solution u of the Riccati Equation

We define $P(t) := \int_{\tau}^{t} \frac{u^2(s)\Delta s}{r(t)+u(t)\mu^*(t)}$. The following conditions are independent of τ.

(P_1) $lim\, \mu^{-1}(t,\tau)\int_{\tau}^{t} P(s)\Delta s$ exists and is finite. Also lim $u(t) = 0$.

(P_2) $lim\, \mu^{-1}(t,\tau)\int_{\tau}^{t} P(s)\Delta s$ exists and is finite. Moreover, $lim\, \mu^{-1}(t,\tau)\int_{\tau}^{t} u(s)\Delta s = 0$.

(P_3) $lim\, P(t)$ exists and is finite. An equivalent condition is: $lim\, \mu^{-1}(t,\tau)\int_{\tau}^{t} P(s)\Delta s$ exists and is finite.

Now we can formulate our main theorem on Riccati techniques for oscillation tests in the following compact and clear form.

Theorem 4.7.6 (Riccati Oscillation Tests): *Let* $u : \mathbb{T}_\tau \to R$ *be a function satisfying condition* (U_1) *with* $(P_1), (P_2)$ *and* (P_3) *defined relative to this function* u. *We then have the following implications.*

(i) *In general,* $(P_3) \Leftarrow (P_2) \Leftarrow (P_1) \Rightarrow (Q_1) \Rightarrow (Q_2) \Rightarrow (Q_3) \Rightarrow (Q_4) \Rightarrow (Q_5)$ *and* $(P_2) \Rightarrow (Q_2)$.

(ii) *If* (R_4), *then* $(P_3) \Rightarrow (P_2)$.

(iii) *If* (R_1), *then* $(P_3) \Rightarrow (P_1)$.

(iv) *If* (R_2) *and* (Q_7), *then* $(P_3) \Leftrightarrow (P_2) \Leftrightarrow (Q_2) \Leftrightarrow (Q_3)$.

(v) *If* (R_1) *and* (Q_7), *then* $(P_3) \Leftrightarrow (P_1) \Leftrightarrow (Q_1) \Leftrightarrow (Q_3)$.

(vi) *If* (S_4), *then* $(P_3) \Leftrightarrow (P_2) \Leftrightarrow (Q_2) \Leftrightarrow (Q_3)$.

(vii) *If* (S_3), *then* $(P_3) \Leftrightarrow (P_2) \Leftrightarrow (Q_2) \Leftrightarrow (Q_4)$.

(viii) *If* (S_2), *then* $(P_3) \Leftrightarrow (P_1) \Leftrightarrow (Q_1) \Leftrightarrow (Q_4)$.

(ix) *If* (S_1), *then* $(P_3) \Leftrightarrow (P_1) \Leftrightarrow (Q_1) \Leftrightarrow (Q_5)$.

(x) *If* (S_5), *then* $\int_\tau^t q(s)\Delta s < \frac{r(\tau)}{\mu^*(\tau)} + \frac{r(t)}{\mu^*(t)}$.

Remarks: (a) In case $\delta^* = 0$ (vanishing graininess) conditions (Q_7) and $(R_2/\text{second part})$ are automatically fulfilled. In this case the statement in (iv) is Theorem (A) in [17] . In (vi) we have an analogous equivalence under the different assumption (S_4). The proofs of (iv) and

(vi) are different and cannot be given in a unified manner(cf. step (l) and (m) in the proof below).

(b) (vi)-(x) are generalizations of various known results, to the case of nonconstant nonvanishing graininess. In the proof we will indicate why the assumption of non-vanishing graininess is crucial for the validity of these results.

(c) In the theorem we have the general assumption (U_1). So if one of the statements is not valid for a pair of functions r and q, we see that (U_1) cannot be fulfilled. As a consequence, (4.7.16) must be oscillatory. Thus Theorem 4.7.6 yields a number of oscillation tests. As an example we get the following oscillation criterion.

Corollary 4.7.4: *If* (R_1) *and* (Q_6) *are valid, then* (4.7.16) *is oscillatory.*

Proof: If (4.7.16) would be nonoscillatory then by (v) of the above theorem we have the following implication, $(U_1), (R_1)$ and $(Q_6) \Rightarrow$ $(U_1), (R_1), (Q_7)$ and $(Q_3) \overset{(v)}{\Rightarrow} (Q_1) \Rightarrow (Q_6)$ cannot hold.

This is a contradiction.

Proof: (of Theorem 4.7.6)(v) is a consequence of (i), (iii) and (iv). (viii) is a consequence of (i), (iii) and (vii).

(a) For convenience we set

$$p(t) = P[u](t) = \frac{u^2(t)}{r(t)+u(t)\mu^*(t)} \geq 0,\ P(t) := \int_\tau^t p(s)\Delta s. \tag{4.7.41}$$

(b) From integrating the Riccati equation (4.7.34) in (U_1) we get,

$$u(t) + P(t) + Q(t) = u(\tau),\ t \geq \tau \tag{4.7.42}$$

and thus the implication $(P_1) \Rightarrow (Q_1)$ holds.

Averaging of this equation yields the existence of

$$lim\, \mu^{-1}(t,\tau)\int_\tau^t [u(s) + P(s) + Q(s)]\Delta s$$

and therefore the implication $(P_2) \Rightarrow (Q_2)$. Because all other implications in (i) are trivial, we have proved (i).

(c) The coupling of u and P in (4.7.41) allows us to give an estimate for u in terms of P and R. We prove the following implication. If for some $M_1, M_2 \geq 0$, we have

$$\int_\tau^t u(s)\mu^*(s)\Delta s \leq M_1 \cdot \int_\tau^t |u(s)|\Delta s + M_2,$$

then

$$\int_\tau^t |u(s)|\Delta s \leq M_1 P(t) + \sqrt{P(t)[R(t) + M_2]}. \tag{4.7.43}$$

By Cauchy's inequality we have

$$\left(\int_\tau^t |u(s)|\Delta s\right)^2 = \left(\int_\tau^t \sqrt{p(s)[r(s)+u(s)\mu^*(s)]}\Delta s\right)^2$$

$$\leq P(t)\cdot[R(t)+\int_\tau^t u(s)\mu^*(s)\Delta s]$$

$$= P(t)\cdot[R(t)+M_2+M_1\int_\tau^t |u(s)|\Delta s].$$

Solving this quadratic inequality and using the property $\sqrt{a+b} \leq \sqrt{a}+\sqrt{b}$ for $a,b\geq 0$ yields

$$\int_\tau^t |u(s)|\Delta s \leq \tfrac{1}{2}\left(M_1P(t)+\sqrt{[M_1P(t)]^2+4P(t)[R(t)+M_2]}\right)$$

$$\leq M_1P(t)+\sqrt{P(t)[R(t)+M_2]}.$$

(d) The "crucial inequality" (4.7.35) appearing in (U_1) can be written as

$$r(t)\delta(t) + u(t)\mu^*(t) \geq 0.$$

Therefore by (4.7.42) we have

$$[P(t)+Q(t)]\mu^*(t) = u(\tau)\mu^*(t) - [r(t)\delta^*(t) + u(t)\mu^*(t)] + r(t)\delta^*(t)$$

$$\leq u(\tau)\mu^*(t) + r(t)\delta^*(t). \tag{4.7.44}$$

We present this simple inequality separately, because it will be used in several steps of this proof.

(e) We are going to show (ii) : $[(R_4)$ and $(P_3)] \Rightarrow (P_2)$.

From the implication (4.7.13) (we set $M_1 = \overline{\mu} := sup\, \mu^*(t)$, $M_2 = 0$, $M_3 = sup\, P(t)$) we see that $\int_\tau^t |u(s)|\Delta s \leq M_1 M_3 + \sqrt{M_3 \cdot R(t)}$.

Therefore (R_4) implies the second condition in (P_2).

(f) Since the outer condition in (iv), (vi), (vii) all imply condition (R_4) in (ii) and the outer condition in (ix) implies (R_1) in (iii) (observe that $r(s) \leq \overline{\mu} \cdot r(s)/\mu^*(s)$, if $\mu^*(s) > 0$), we can complete the proof, if we show the implications $(Q_3) \Rightarrow (P_3)$, if [(R_2) and (Q_7)] (for (iv)), $(Q_3) \Rightarrow (P_3)$, if (S_4)(for (vi)), $(Q_4) \Rightarrow (P_3)$, if (S_3)(for (vii)), $(Q_5) \Rightarrow (P_3)$, if (S_1) (for (ix)).

(g) To show (ix), By (U_1) : (4.7.35) we have

$$0 = u(t) - u(\tau) + Q(t) + P(t) \geq -\frac{r(t)}{\mu^*(t)} - u(\tau) + Q(t) + P(t).$$

By (S_1) and (Q_5), P must be bounded. So the limit of P exists and is finite.

(h) From now on, in order to show the first three lines in the list in (f) we assume to the contrary that (P_3) does not hold.

$$lim\, P(t) = \infty \text{ or (equivalently) } \mu^{-1}(t,\tau)\int_\tau^t p(s)\Delta s = \infty. \quad (4.7.45)$$

(j) We show (vii). From (S_3) and (Q_4) we derive a contradiction by dividing equation (4.7.44) by $\mu^*(t)$, averaging

$$\mu^{-1}(t,\tau)\int_\tau^t [P(s)+Q(s)]\Delta s \le u(\tau)+\mu^{-1}(t,\tau)\int_\tau^t \frac{r(s)}{\mu^*(s)}\Delta s$$

and then apply $limsup$.

(k) Preparation for (iv) and (vi). From equation (4.7.42), condition (Q_3) and the contradictory assumption (4.7.45) we conclude,

$$\begin{aligned}
&- \mu^{-1}(t,\tau)\int_\tau^t u(s)\Delta s\\
&= \mu^{-1}(t,\tau)\int_\tau^t [Q(s)-u(\tau)]\Delta s+\mu^{-1}(t,\tau)\int_\tau^t P(s)\Delta s\\
&\ge M_4+\mu^{-1}(t,\tau)\int_\tau^t P(s)\Delta s \ge \tfrac{1}{2}\mu^{-1}(t,\tau)\int_\tau^t P(s)\Delta s
\end{aligned}$$

(for t large) and therefore

$$\int_\tau^t |u(s)|\Delta s \ge -\int_\tau^t u(s)\Delta s \ge \tfrac{1}{2}\int_\tau^t P(s)\Delta s \to \infty \qquad (4.7.46)$$

for large t. Via (4.7.46) we can now reformulate our implication (4.7.43) as follows. If

$$\int_\tau^t u(s)\mu^*(s)\Delta s \le M_1\cdot\int_\tau^t |u(s)|\Delta s + M_2.$$

then

$$\int_\tau^t P(s)\Delta s \le M_1P(t)+\sqrt{P(t)[R(t)+M_2]}. \qquad (4.7.47)$$

(l) For the proof of (iv) let (R_2) and (Q_7) hold. From inequality (4.7.44) we see that

$$sup\, P(t)\,\mu^*(t)<\infty \qquad (4.7.48)$$

Using (4.7.42),(4.7.45) and (Q_7) we estimate,

$$\int_\tau^t u(s)\mu^*(s)\Delta s = \int_\tau^t [\,u(s) - P(s) - Q(s)\,]\mu^*(s)\Delta s$$

$$\leq \int_\tau^t [\,u(s) - P(s) + M_5\,]\mu^*(s)\Delta s \leq M_2 < \infty.$$

Now, from the implication (4.7.47) (see $M_1 = 0$) and (R_2) we obtain

$$\left(\int_\tau^t P(s)\Delta s\right)^2 \leq 4P(t)\cdot[\,R(t) + M_2] \leq 4P(t)\cdot M_6\,\mu(t,\tau).$$

Setting $W(t) = \int_\tau^t P(s)\Delta s$, we can conclude

$$\int_{\tau_2}^t \frac{\Delta s}{\mu(s,\tau)} \leq 4\cdot M_6\cdot \int_{\tau_2}^t \frac{W^\Delta(s)}{W^2(s)}\Delta s,\; t\geq \tau_2 \geq \tau.$$

By Theorem 4.7.2 the left hand side goes to infinity, the right hand side is bounded because of (4.7.48). This contradiction shows (iv).

(m) Now for the proof of (vi) let (S_4) hold. With $M_1 = \overline{\mu}$, $M_2 = 0$ the condition in (4.7.47) is fulfilled. We get,

$$\tfrac{1}{2}\int_\tau^t P(s)\Delta s \leq \overline{\mu}\;\cdot P(t) + \sqrt{P(t)R(t)}. \qquad (4.7.49)$$

From Theorem 4.7.3 (ii), inequality (4.7.44) and (Q_3) we get

$$lim\;sup\;\mu^{-\frac{1}{2}}(t,\tau)\,P(t) \leq lim\;sup\;\mu^{-\frac{3}{2}}(t,\tau)\;\cdot\int_\tau^t P(s)\Delta s$$

$$\leq lim\;sup\;\mu^{-\frac{3}{2}}(t,\tau)\;\cdot\int_\tau^t[\,u(\tau) - Q(s) + \tfrac{r(s)}{\mu^*(s)}]\Delta s < \infty.$$

Therefore dividing the right hand side in (4.7.49) by $\mu(t,\tau)$ gives

$$\tfrac{1}{2}\mu^{-1}(t,\tau)\int_\tau^t P(s)\Delta s$$

$$\leq \overline{\mu}\;\cdot\mu^{-1}(t,\tau)\,P(t) + \sqrt{\mu^{-\frac{1}{2}}(t,\tau)\,P(t)\cdot\;\mu^{-\frac{3}{2}}(t,\tau)R(t)}$$

$$< \infty\,.$$

which is bounded above. This contradicts (4.7.45).

(n) We directly show (x). Here we can assume that $\tau < t$. Using (4.7.42) and again (4.7.35) we estimate

$$\int_\tau^t q(s)\Delta s = u(\tau) - u(t) - p(\tau)\mu^*(\tau) - \int_{\sigma(\tau)}^t p(s)\Delta s$$

$$< \frac{r(t)}{\mu^*(t)} + u(\tau) - \frac{u^2(\tau)\mu(\tau)}{r(\tau)+u(\tau)\mu^*(\tau)}$$

$$= \frac{r(t)}{\mu^*(t)} + \frac{r(\tau)}{\mu^*(\tau)} \cdot \left[\frac{u(\tau)}{\frac{r(\tau)}{\mu^*(\tau)} + u(\tau)}\right]$$

$$< \frac{r(t)}{\mu^*(t)} + \frac{r(\tau)}{\mu^*(\tau)} .$$

4.8 Convexity of Solutions Relative to the Initial Data

Recently [44] a nice result is proved, which has been overlooked in the literature. This is to show that if $f(t,x)$ is convex in x uniformly in t, the solution $x(t,t_0,x_0)$ is also convex in x_0 for each $t > t_0$. In this section, we extend this result to a Banach space in the frame work of time scales.

Let Γ be a real Banach space. Consider

$$x^\Delta = f(t,x), \qquad x(0) = x_0 \tag{4.8.1}$$

where $f \in C_{rd}[\mathbb{T} \times \Omega, \Gamma]$ and $\Omega \subseteq \Gamma$ is a suitable subset.

A proper subset S of a Banach space Γ is said to be a cone if $\rho S \subset S, \rho \geq 0, S+S \subset S, \overline{S} = S$ and $S \cap \{-S\} = 0$, where 0 denotes the null element of the Banach space Γ and $\overline{S}$ denotes the closure of S.It is assumed that S is a distance set, that is, to each point $x \in \Gamma$, there corresponds a point $y \in S$ such that $d(x,S) = \|x-y\|$. In this connection we note that every closed convex set in a reflexive Banach space is a distance set. For details we refer to [44]. The cone S induces the order relations in S defined by

$$x \leq y \quad if\, and\, only\, if \quad y-x \in S, \quad x < y \quad if\, and\, only\, if \quad y-x \in S^0$$

where S^0 denotes the interior of S. Let S^* be the set of all linear functionals l on Γ such that $lx \geq 0$ for all $x \in S$ and let S_0^* be the set of all continuous linear functionals on Γ such that $lx > 0$ for all $x \in S^0$. A function f from Γ into Γ is said to be quasimonotone nondecreasing if $x \leq y$ and $lx = ly$ for some $l \in S_0^*$, then $l(f(x)) \leq l(f(y))$.

At a point $x \in \Gamma$

$$f(t,x+h) = f(t,x) + L(t,x,h) + \|h\|\eta(t,x,h)\,, \quad h \in \Gamma$$

where $L(t,x,.)$ is a linear operator, $\lim_{\|h\|\to 0} \|\eta(t,x,h)\| = 0$. $L(t,x,h)$ is called the Frechet derivative of the function f at x with increment h and $\eta(t,x,h)$ is called the remainder of the derivative. The operator $L(t,x,.) : \Gamma \to \Gamma$ is called the Frechet derivative of f at x and is denoted by $f_x(t,x)$.

We need the following lemmas for our subsequent analysis, whose proofs can be modeled on the basis of [35] for time scales.

Lemma 4.8.1: *Let S be a cone in Γ . Assume that*

(i) $u, v \in C'_{rd}[\mathbb{T}, \Gamma]$, $f \in C_{rd}[\mathbb{T} \times \Gamma, \Gamma]$ *and* $f(t,x)$ *is quasimonotone nondecreasing in* x, $f(t,x)\mu^*(t) + x$ *is non decreasing in* x, *relative to* S *for each* $t \in \mathbb{T}$,

(ii) $u^\Delta(t) - f(t,u(t)) \le v^\Delta(t) - f(t,v(t)), t \in \mathbb{T}$,

(iii) $\|f(t,x) - f(t,y)\| \le g(t, \|x-y\|)$, $x \in \Gamma \backslash S$, $y \in \partial S$ *(the boundary on* S), *where* g *is a uniqueness function.*

(iv) S *is a distance set.*

Then $u(t_0) \le v(t_0)$ *implies that* $u(t) \le v(t)$, $t \in \mathbb{T}$.

Lemma 4.8.2: *Let* $f \in C_{rd}[\mathbb{T} \times \Gamma, \Gamma]$ *and suppose that Frechet derivative* $f_x(t,x)$ *of* f *exists for* $x \in B[x_0, b]$. *If* $f(t,x)$ *is convex on* $B[x_0, b]$ *for each* $t \in \mathbb{T}$, *then* $\eta(t,x,h) \in S$, *that is,*

$$f(t,x+h) - f(t,x) \geq f_x(t,x)h$$

where S *is a cone in* Γ *and* $B[x_0, b] = \{x : \|x - x_0\| \leq b\}$, $b > 0$.

Lemma 4.8.3: *Let* $f \in C_{rd}[\mathbb{T} \times \Gamma, \Gamma]$ *and let* $f(t,x)$ *be convex on* $B[x_0, b]$ *for each* $t \in \mathbb{T}$. *Suppose that the Frechet derivative* $f_x(t,x)$ *of* f *exists for* $x \in B[x_0, b]$. *For each* $t \in \mathbb{T}$, *if* $f(t,x)$ *quasimonotone nondecreasing in* x, $[f_x(t,y)\mu^*(t) + I]v$ *is nondecreasing in* v, *then the function* $G(t,v) = f_x(t, \alpha(t))v$ *quasimonotone nondecreasing in* v *for each* $t \in \mathbb{T}$, *where* $\alpha \in C_{rd}[\mathbb{T}, B[x_0, b]]$.

We are now in a position to prove our main result.

Theorem 4.8.1: *Assume that*

(i) $f \in C_{rd}[\mathbb{T} \times \Gamma, \Gamma]$, $f_x(t,x)$ *exists and is rd-continuous for* $(t,x) \in \mathbb{T} \times B[x_0, b]$, *and there exists* M *and* L *such that* $\|f(t,x)\| \leq M$ *and* $\|f(t,x)\| \leq L$, *for* $(t,x) \in \mathbb{T} \times B[x_0, b]$ *where* $a(M + Lb) \leq b$.

(ii) *for each* $t \in \mathbb{T}$, $f(t,x)$ *is convex on* $B[x_0, b]$

(iii) *for each* $f(t,x)$ *quasimonotone nondecreasing in* x,

$f_x(t,y)\mu^*(t)+x$ *is nondecreasing in* x,

Then the solution $x(t,x_0)$ *of (4.8.1) exists in all of* $\mathbb{T}$ *and* $x_0 \to x(t,x_0)$ *is convex on a convex set* $C \subseteq \Gamma$.

Proof: By (i) it is easy to conclude that $f(t,x)$ satisfies a Lipschitz condition in x for $(t,x) \in \mathbb{T} \times B[x_0, b]$ and consequently there exists unique solutions $x(t,x_0)$ for (4.8.1) on $\mathbb{T}$, where x_0 varies in a convex set $C \subseteq \Gamma$.

Let $y(t) = x(t; p(s)),\ \ z(t) = x(t; y_0)$

and $u(t) = z(t) - y(t)$ where $p(s) = sx_0 + (1-s)y_0$, $0 \leq s \leq 1$

Then by Lemma 4.8.2, we have

$$\begin{aligned} u^\Delta &= z^\Delta - y^\Delta \\ &= f(t,z) - f(t,y) \\ &\geq f_x(t,y)\,u \end{aligned}$$

and

$$\begin{aligned} u(t_0) &= z(t_0) - y(t_0) \\ &= y_0 - [s\,x_0 + (1-s)\,y_0] \\ &= -\,s\,[x_0 - y_0]. \end{aligned}$$

So we get

$$u^\Delta \geq f_x(t,y)\,u\,,\ u(t_0) = -\,s\,[x_0 - y_0]$$

Consider the IVP

$$w^{\Delta} \geq f_x(t,y)\,w,\; w(t_0) = I$$

which implies , by setting

$$v(t) = \; - \, w(t)\, s[\,x_0 - y_0],$$

that

$$v^{\Delta} = f_x(t,y)\,v\,,\; v(t_0) = \; - \, s\,[\,x_0 - y_0].$$

Since $f_x(t,y)\,v$ is quasimonotone nondecreasing in v (lemma 4.8.1), by lemma 4.8.2, we get

$$u(t) \geq v(t),\;\; t \in \mathbb{T},$$

which is equivalent to

$$x(t;p(s)) \leq x(t;y_0) + w(t)\,s\,(x_0 - y_0). \tag{4.8.2}$$

Now let

$$y(t) = x(t;p(s)),\;\; z(t) = x(t;x_0)$$

and

$$u(t) = z(t) - y(t).$$

By lemma 4.8.3, we have

$$u^{\Delta} \geq f_x(t,y)\,u\,,\; u(t_0) = (1 - s\,)[\,x_0 - y_0]$$

Therefore, as before,

$$u(t) \geq w(t)\,(1-s)[\,x_0 - y_0],\ t \in \mathbb{T},$$

that is,

$$x(t;p(s)) \leq x(t;y_0) - w(t)\,(1-s\,)(x_0 - y_0). \qquad (4.8.3)$$

Multiplying (4.8.2) by $(1-s)$, (4.8.3) multiplied by s and adding, we get

$$x(t;p(s)) \leq\ sx(t;x_0) + (1-s\,)x(t;y_0),$$

which completes the proof.

4.9 Invariance Principle

In this section, we shall consider the extension of the popular invariance principle of LaSalle [45, 46] in the present framework of time scales.

Consider the initial value problem for an autonomous system on a time scale $\mathbb{T}$

$$x^{\Delta} = f(x) \quad x(0) = x_0, \qquad (4.9.1)$$

where $f \in C[G^*, \mathbb{R}^n]$ and G^* is an open set in $\mathbb{R}^n$. We assume for convenience that the solutions of (4.9.1) in G^*, denoted by $x(t, x_0)$, are unique. Note that when $\mathbb{T} = \mathbb{R}$ (4.9.1) coincides with the initial value problem

$$\frac{dx}{dt} = f(x), \quad x(0) = x_0,$$

whose solutions are unique in both directions of time, continuous and can go to infinity in finite time, whereas when $\mathbb{T} = \mathbb{Z}$ (4.9.1) coincides with the difference equation

$$x(n+1) - x(n) = f(x_n) \quad x(0) = x_0$$

where $n \in N_0$. The solution to the initial value problem is $x(n) = T^n x_0$ where T^n is the nth iterate of T: $T^{n+1} = T(T^n)$ and $T^0 = I$, the identity map. The motion from x_0 is the sequence of states $x_0, Tx_0, T^2x_0, \ldots, T^n x_0, \ldots$. In contrast to differential equations, the solutions to differential equations are in general, not uniquely determined in backward time and an additional condition which guarantees unique backward continuation of solutions has to be imposed. The condition is known as the regressivity condition.

Definition 4.9.1: The mapping $f : T^* \times X \to X$ (a Banach space) is called regressive in $t \in T^*$ if the mapping

$$I + f(t, \cdot)\mu^*(t) : \ X \to X$$

where I is the identity map is invertible. f is called regressive (on $\mathbb{T}$) if it is regressive in each $t \in \mathbb{T}^*$. One can verify that a mapping $f : \mathbb{T}^* \times X \to X$ which fulfills the condition

$$|f(t, x_1) - f(t, x_2)| \leq L|x_1 - x_2|, \quad x_1, x_2 \in X$$

for some $t \in \mathbb{T}^*$, $L.\mu^*(t) < 1$, $L > 0$, is regressive.

The concepts of positive and negative limit sets play an important role in the extension of Lyapunov's stability theory since information about the location of the limit set of a motion is also information about its asymptotic behavior. In his Invariance principle, LaSalle showed, by exploiting the invariance property of limit sets, that, suitably defined, Lyapunov functions give information about the location of limit sets. In this section, we show that the Invariance Principle holds true on arbitrary time scales. Following LaSalle, we give the following definitions.

Definition 4.9.2: Limit Sets of Motions.

Let $\phi : (\alpha, \omega) \cap \mathbb{T} \to G^*$ where $\alpha, \omega \in \mathbb{T}$, $-\infty \leq \alpha < 0 < \omega \leq -\infty$. A point p is said to be a *positive* (*negative*) limit point of ϕ if there exists a sequence of points $t_n \in (\alpha, \omega) \cap \mathbb{T}$ such that $t_n \to \omega$ ($t_n \to \alpha$) as $n \to \infty$ and $\phi(t_n) = p$. The set of all positive (negative) limit points of ϕ, $\Omega(\phi)$ $(A(\phi))$ is called the positive (negative) limit set of ϕ. The interval $(\alpha, \omega) \cap \mathbb{T}$ is said to be *maximal* relative to G^* is empty and if $-\infty < \alpha$ implies that $A(\phi) \cap G^*$ is empty.

Basic properties of solutions: For each $t \in \mathbb{T}$ define $T(t) : \mathbb{R}^n \to \mathbb{R}^n$ such that

$$x(t,\, x_0) = T(t)\, x_0$$

$$x(0, x_0) = x_0 = T(0)x_0 \quad T(0) = I,$$

then, from the basic theory of differential equations on time scale, we have the following properties.

(1) Each solution $x(t, x_0)$ of (4.9.1) satisfying $x(0) = x_0$ has for each $x_0 \in G^*$ a maximal interval of definition $I(x_0) = (\alpha(x_0), \omega(x_0)) \cap \mathbb{T}$, $-\infty \leq \alpha < 0 < \omega \leq +\infty$.

(2) Let $s, t \in \mathbb{T}$ such that $s \in I(x_0)$ and $t \in I(x(s, t_0))$, then one can show that $\sigma(t+s) \in I(x_0)$ and $T(\sigma(s+t))x_0 = T(\sigma(s)) \cdot T(t)x_0$.

(3) $x \in \mathbb{T}^* \times G^* \to \mathbb{R}^n$ is rd-continuous, i.e.,

(i) it is continuous at each (t, x) with right-dense or maximal t;

(ii) for $t_n < t$, $t_n \in I(x(t_n, x_0))$, $(t_n, x(t_n, x_0)) \to (t, y) \in I(y) \times G^*$ then $x(t_n, x(t_n, x_0)) \to x(t, y)$ at each left-dense point.

Definition 4.9.3: Relative to (4.9.1), a set $H \subset \mathbb{R}^n$ is said to be *positively* (*negatively*) weakly *invariant* if $x_0 \in H \cap G^*$ implies that $T(t)x_0 \in H$ for all $t \in [0, \omega(x_0)) \cap \mathbb{T}\,(t \in (\alpha(x_0), 0])$. H is said to be *weakly invariant* if it is positively and negatively invariant. If, in addition, $I(x_0) = \mathbb{T}$ for each $x_0 \in H \cap G^*$, H is said to be invariant.

Definition 4.9.4: A solution $x(t, x_0) = T(t)x_0$ is said to be *positively* (*negatively*) *precompact* relative to G^* if it is bounded for all $t \in [0, \omega(x_0)) \cap \mathbb{T}$ $(t \in (\alpha(x_0), 0] \cap \mathbb{T})$ and has no limit points on the boundary of G^*.

Note: In case $\mathbb{T} = \mathbb{Z}$, $x(t, x_0) = T(t)x_0$ is precompact but when $\mathbb{T} = \mathbb{R}$ the solution can go to infinity in finite time but, since in both cases $I(x_0)$ is maximal, the precompactness of $X(t, x_0)$ implies that $\omega(x_0) = +\infty$. In addition, if a set H is precompact relative to G^* and weakly invariant, then it is invariant.

Definition 4.9.5: A closed invariant set is said to be *invariant connected* if it is not the union of two nonempty disjoint closed invariant sets.

In case $\mathbb{T} = \mathbb{R}$, invariantly connected means connectedness.

Basic Properties of Limit Sets:

We consider only positive limit sets. Similar results can be obtained for negative limit sets by substituting $-t$ for t.

Theorem 4.9.1: *Every positive limit set $\Omega(x_0)$ is closed and positively invariant.*

Proof: Each positive limit set $\Omega(x_0)$ is closed for if the complement, $\Omega^c(x_0)$, of $\Omega(x_0)$ is closed then $\Omega^c(x_0)$ contains its limit points. However, the limit points are all in $\Omega(x_0)$. To show positive invariance of $\Omega(x_0)$, let $y \in \Omega(x_0) \cap G^*$ and $t \in I(y)$. Since $\Omega(x_0) \cap G^*$ is nonempty, $\omega(x_0) = \infty$. Then there is a sequence $t_n \in I(x_0)$ such that $t_n \to \infty$ and $x(t_n, x_0) \to y$ as $n \to \infty$. For n sufficiently large $I(y) \cap I(x(t_n, x_0))$ is nonempty and $x(t, x(t_n, x_0)) = x(t + t_n, x_0) \to x(t, y)$. Hence $x(t, y) \in \Omega(x_0)$, and $\Omega(x_0)$ is positively invariant.

Theorem 4.9.2: *If $x(t, x_0)$ is positively precompact, then $\Omega(x_0)$ is in G^*, and is nonempty, compact, invariant, invariantly connected, and is the smallest closed set that $x(t, x_0)$ approaches as $t \to \infty$.*

Proof: Since $x(t, x_0) = T(t)x_0$ is positively precompact, it is bounded and this in turn implies that $\Omega(x_0)$ is nonempty bounded and in G^*, and since it is closed it is also compact. Since it is positively invariant it is invariant.

We show next that $x(t, x_0) = T(t)x_0 \to \Omega(x_0)$ when $T(t)x_0$ is precompact. Since $T(t)x_0$ and $\Omega(x_0)$ are both bounded, the distance between them, denoted here by $\rho(T(t)x_0,\Omega(x_0))$, is bounded and we conclude that if $T(t)x_0$ does not approach $\Omega(x_0)$, then there is a sequence $t_n \in I(x_0)$ such that $t_n \to \infty$, and $x(t_n, x_0) = T(t_n)x_0$ converges but $\rho(T(t_n)x_0, \Omega(x_0))$ does not approach 0 as $t_n \to \infty$. This is a contradiction since the limit of $x(t_n, x_0)$ is in $\Omega(x_0)$. Hence, $x(t_n, x_0) \to \Omega(x_0)$ as $t_n \to \infty$. If $T(t_n)x_0 \to E$ as $t_n \to \infty$ and E is a closed set, then $\Omega(x_0) \subset E$. Hence, $\Omega(x_0)$ is the smallest closed set that $T(t_n)x_0$ approaches as $t_n \to \infty$.

To show that $\Omega(x_0)$ is invariantly connected assume that $\Omega(x_0)$ is the union of 2 disjoint closed nonempty invariant sets Ω_1 and Ω_2. Since $\Omega(x_0)$ is compact, so are Ω_1 and Ω_2. Then there exists disjoint open sets U_1 and U_2 such that $\Omega_1 \subset U_1$ and $\Omega_2 \subset U_2$. Also, since for each t, $T(t)$ is continuous on Ω_1, and therefore uniformly continuous on Ω_1, there exists an open set V_1 such that $\Omega_1 \subset V_1$ and $T(t)(V_1) \subset U_1$. Since

$\Omega(x_0)$ is the smallest closed set that $T(t)x_0$ approaches, there must exists a sequence $t_n \in I(x_0)$ such that $t_n \to \infty$ as $n \to \infty$ such that $T(t_n)x_0$ intersects both V_1 and U_2 an infinite number of times. But this implies the existence of a subseqeunce t_{n_i} such that $T(t_{n_i})x_0$ is neither in V_1 nor in U_2. But this is a contradiction, and $\Omega(x_0)$ is invariantly connected.

Definition 4.9.6: Let $V : C[G^*, \mathbb{R}]$ and let G be a subset of G^*. V is said to be a Lyapunov function of (4.9.1) on G if

(i) V is continuous,

(ii) $\dot{V}(x) = V_x(x).f(x) \leq 0$.

Relative to a Lyapunov function of (4.9.1) on G, we introduce the following notation.

$$E = \{x : \dot{V}(x) = 0,\ x \in \overline{G} \cap G^*\}.$$

M is the largest invariant set in E.

M^* is the largest weakly invariant set in E.

M^+ is the largest positively invariant set in E.

If M^* is compact, then $M = M^*$. We also have $M \subset M^* \subset M^+$.

$$V^{-1}(c) = \{x \in \overline{G} \cap G^* : V(x) = c\}.$$

We shall now show that suitably defined Lyapunov functions can give information about the location of the limit set of a motion.

Theorem 4.9.3: (Invariance Principle) *Let V be a Lyapunov function of* (4.9.1) *on G, and let $x(t, x_0)$ be a solution of* (4.9.1) *that remains in G for all $t \in [0, \omega(x_0)) \cap \mathbb{T}$. Then for some c, $\Omega(x_0) \cap G^* \subset M^* \cap V^{-1}(c)$. If $x(t, x_0)$ is precompact then $x(t) \to M \cap V^{-1}(c)$.*

Proof: Assume that $y \in \Omega(x_0)$. Then $\omega(x_0) = \infty$ and there exists a sequence $t_n \in I(x_0)$ such that $x(t_n, x_0) \to y$. Since $V(x(t, x_0))$ is nonincreasing with respect to t, $V(x(t, x_0)) \geq V(y)$ for all $t \in [0, \infty) \cap \mathbb{T}$, and $V(x(t, x_0)) \to c = V(y)$ for each $y \in \Omega(x_0)$. Since $\Omega(x_0)$ is weakly positively invariant, $\Omega(x_0) \subset E$ and therefore $\Omega(x_0) \subset M^*$. Hence, $\Omega(x_0) \cap G^* \subset M^* \cap V^{-1}(c)$. If $x(t, x_0)$ is precompact, then $\Omega(x_0)$ is invariant and $\Omega(x_0) \subset M \cap V^{-1}(c)$. Since $x(t, x_0) \to \Omega(x_0)$ as $t \to \infty$, $x(t) \to M \cap V^{-1}(c)$ as $t \to \infty$.

4.10 Notes

The results of Section 4.1 concerning the method of lower and upper solutions, monotone iterative technic for initial and periodic boundary value problems are adapted from the work of Kaymakcalan [22], and Sivasundaram [54]. See Ladde, Lakshmikantham and Vatsala [29] for corresponding results relative to continuous dynamic systems. Refer to Leela and Sivasundaram [56], and Leela and Vatsala [48], for the contents of Sec 4.2 with respect to the method of quasilinearization. See also Bellman [7], Bellman and Kalaba [8], Lakshmikantham and Malek [35], and Lakshmikantham [31] for recent trends in the method of quasilinearization. The results of Section 4.3, dealing with monotone flows and fixed point theorems in finite and infinite dimensions are taken from Lakshmikantham and Kayamakcalan [33] and Lakshmikantham [32]. See also Gouze and Hadeler [16] for related results. The work relative to invariant manifolds contained in Sec 4.4 is taken from Aulbach and Hilger [6]. See for details Hilger [18, 19]. For corresponding results on continuous dynamic systems see Palmer [52] and Aulbach [3, 4]. We refer to Drici [13], Drici and Lakshmikantham [14] for work in Sec 4.5 relative to uncertain dynamic systems and practical stability. See Corless and Leitmann [10] for the survey of uncertain dynamic systems. The results on two point boundary value problems given in Sec 4.6 are taken from Murthy and Rao [50]. For the work relative to Sturmian theory described in Section 4.7, see Erbe and Hilger [15]. Section 4.8, covers the work of Sivasundaram [55]. See Lakshmikantham, Shahzad and Walter [44] for corresponding results for continuous systems in finite dimensions. Finally, in Section 4.9, we develop the invariance principle of LaSalle for dynamic systems on time scales. See LaSalle [45, 46].

REFERENCES

[1] Agarwal, R.P., *Difference Equations and Inequalities*, Marcel Dekker, New York 1992.

[2] Aulbach, B., *Analysis auf Zeitmengen*, Lecture Notes, Universitat Augsburg 1990.

[3] Aulbach, B., *Continuous and Discrete Dynamics near Manifolds of Equilibria*, Lecture Notes in Math. **1058**, Springer, Berlin, Heidelberg, New York, Tokyo 1984.

[4] Aulbach, B., Hierarchies of invariant manifolds, *J. Nigerian Math. Soc.* (to appear).

[5] Aulbach, B. and Hilger, S., *Linear Dynamic Processes with Inhomogeneous Time Scales*, Nonlinear dynamics and Quantum Dynamical Systems, Akademie Verlag, Berlin 1990.

[6] Aulbach, B. and Hilger, S., *A Unified Approach to Continuous and Discrete Dynamics* **53**, Qualitative Theory of Differential Equations, Szeged, Hungary 1988.

[7] Bellman, R., *Methods of Nonlinear Analysis*, Vol. II, Academic Press, New York 1973.

[8] Bellman, R. and Kalaba, R., *Quasilinearization and Nonlinear Boundary Value Problems*, American Elsevier, New York 1965.

[9] Christiansen, F.B. and Fenchel, T.M., *Theories of Populations in Biological Communities*, Springer, Berlin 1977.

[10] Corless, M. and Leitmann, G., Deterministic Control of Uncertain systems: A Lyapunov Theory Approach, Chapter II In: *Deterministic Control of Uncertain Systems*, Edited by A.S.I. Zinober, Peter Peregrinus, London 1990.

[11] Dieudonne, J., *Foundations of Modern Analysis*, Prentice Hall, New Jersey 1966.

[12] Drici, Z., Stability theory of large scale nonlinear dynamic systems, Ph.D. Dissertation, Florida Institute of Technology, June 1994.

[13] Drici, Z., Invariance principle for dynamic systems on time scales, (to appear).

[14] Drici, Z. and Lakshmikantham, V., Stability of conditionally invariant sets and controlled uncertain systems on time scales, *Math Problems in Engineering* **1** (1995), 1-10.

[15] Erbe, L. and Hilger, S., Sturmian Theory on measure Chains, *Differential equations and Dynamical systems* **1**, No. 3 (1993) 223-246.

[16] Gouze, J.L. and Hadeler, K.P., Monotone flows and ordered intervals, *Nonlinear World* **1** (1994), 23 -34.

[17] Hartman, P., Difference equations: discongacy, principal solution, Green's function, complete monotonicity, *Trans. Amer.. Math. Soc.* **246** (1978), 1 -30.

[18] Hilger, S., Ein Malskettenkalul mit Anwendung auf Zentrums mannigfaltigkeiten, Ph.D. Dissertation, Univ. Wuzburg 198

[19] Hilger, S., Analysis on measure chains A unified approach to continuous and discrete calculus, *Res. in Math.* **18** (1990), 18 56.

[20] Hilger, S., Smoothness of invariant manifolds, (to appear).

[21] Kalman, R.E. and Bertram, J.E., Control system analysis and design via the "second method of Lyapunov", I. Continuous time systems, II. Discrete time systems, *Trans. AASME Ser. D. J. Basic Engineering* (1960), 371-393, 394-400.

[22] Kaymakcalan, B., Existence and comparison results for dynamic systems on time scale, *JMMA* **172** (1993), 243-255.

[23] Kaymakcalan, B., Lyapunov stability theory for dynamic systems on time scales, *JAMSA* **S:3** (1993), 243-2ss.

[24] Kaymakcalan, B., Monotone iterative method for dynamic systems on time scales, *Dyn. Sys. and Appl.* **2**:2 (1993), 213-220.

[25] Kaymakcalan, B., Stability analysis in terms of two measures for dynamic systems on time scales, *JAMSA* **6** (1993), 325-344.

[26] Kaymakcalan, B. and Leela, S., A survey of dynamic systems on time scales, *Nonlinear Times and Digest* **1** (1994), 37-60.

[27] Kaymakcalan, B. and Rangarajan, L., Variation of Lyapunov's method for dynamics systems on time scales, *JMAA* **185** (1994), 356-366.

[28] Kocic, V.L.J. and Ladas, A., *Global Behavior of Nonlinear Difference Equations of Higher Order*, Kluwer Publishers 1993.

[29] Ladde, S.G, Lakshmikantham, V and Vatsala A.S., *Monotone Iterative Technics For Nonlinear Differential equations*. Pitman, Boston, 1985.

[30] Lakshmikantham, V., Further improvement of generalized quasi-linearization, *Nonlinear Analysis* (to appear).

[31] Lakshmikantham, V., Generalized quasilinearization and rapid convergence, In: *Proc. of Conference on Neural, Parallel, and Scientific Computations*, Atlanta, GA (1995), (to appear).

[32] Lakshmikantham, V., Monotone flows and fixed points for dynamic systems on time scales in a Banach space, *Appl. Analysis* **56** (1995), 175-184.

[33] Lakshmikantham, V. and Kaymakcalan, B., Monotone flows and fixed points for dynamic systems on time scales, *Comp. Math. Appl.* **28** (1994), 185-189.

[34] Lakshmikantham, V. and Leela, S., *Differential and Integral Inequalities*, Vol.I, Academic Press, New York 1969.

[35] Lakshmikantham, V. and Leela, S., *Nonlinear Differential Equations in Abstract Spaces*, Pergamon Press, Oxford 1981.

[36] Lakshmikantham, V. and Leela, S., Controlled uncertain dynamic systems and stability of moving invariant sets, *Problems in Nonlinear Analysis in Engineering Applications*, Kazan, Russia (to appear).

[37] Lakshmikantham, V. and Liu, X., *Stability Analysis in Terms of Two Measures*, World Scientific Publishers, Singapore 1993.

[38] Lakshmikantham, V. and Malek, S., Generalized quasilinearization, *Nonlinear World* **1** (1994), 59-63.

[39] Lakshmikantham, V. and Sivasundaram, S., Stability of moving invariant sets and uncertain dynamic systems on time scales, *Proc. Nonlinear Problems in Aviation and Aerospace* (to appear).

[40] Lakshmikantham, V. and Trigiante, D., *Theory of Difference Equations: Numerical Methods and Applications*, Academic Press, New York 1988.

[41] Lakshmikantham, V. and Vatsala, A.S., Stability of moving invariant sets, *Volume dedicated to Professor S. Leela*, Gordon and Breach, London 1996 (to appear).

[42] Lakshmikantham, V., Leela, S., and Martynyuk, A.A., *Stability Analysis for Nonlinear Systems*, Marcel Dekker, New York 1989.

[43] Lakshmikantham, V., Shahzad, N., and Sivasundaram, S., Nonlinear variation of parameters formula for dynamic systems on measure chains, *Dynamics of Continuous, Discrete and Impulsive Systems* **1** (1995), 255-265.

[44] Lakshmikantham, V., Shahzad, N., and Walter,W., Convex dependence of solutions of differential equations in a Banach space relative to initial data, *Nonlinear Analysis* (to appear).

[45] LaSalle, J.P., *Stability of dynamical systems*, SIAM, Philadelphia, PA 1976.

[46] LaSalle, J.P. Stability for difference equations, In: *Studies in Ordinary Differential Equations*, MAA Studies in Math.

[47] Leela, S. and Shahzad, N., On stability of moving conditionally invariant sets, *Nonlinear Analysis*, (to appear).

[48] Leela, S. and Vatsala, A.S., Dynamic Systems on time scales and Generalized Quasilinearization, (to appear).

[49] Leitmann, G., One approach to the control of uncertain dynamic systems, *Appl. Math. Comp.* **70** (1995), 261-272.

[50] Murthy, K.N. and Rao, Y. S., Two point boundary value problems in homogeneous time scale linear dynamic process, *JMAA* **184** (1994), 22-34.

[51] Otero, D., Giuliani, D. and Sassano, M., Temporal dimension and transition out of chaos, *Physica* **A 178** (1991), 280-288.

[52] Palmer, K.J., Qualitative behavior of a system of ODE near an equilibrium point. A generalization of Hartman Grobmantheorem **14** (1977), 1-31.

[53] Siljak, D.D., Ikeda, M. and Ohata, Y., Parametric stability, *Proc. Universita di Genova-Ohio State University First Conference*, Birkhauser (1991), 1-20.

[54] Sivasundaram, S., Method of mixed monotony for dynamic system on time scales, *J. Nonlinear Differential Equations Theory Method and Applications,* **1**,No. 1& 2, (1995) 93-102.

[55] Sivasundaram, S., Convex dependence of solutions of dynamic systems on time scales relative to initial data, *Advances in Nonlinear Dynamics, Gordon Breach Science Publishers, London,*(1996) to appear.

[56] Sivasundaram, S. and Leela, S., Dynamic systems on time scale and superlinear convergence of iterative process, In: *Inequalities and Applications*, World Scientific Publishing Co., Singapore (1994), 431-436.

INDEX

A

B

C

D

E

F

G

I

J

L

M

N

O

P

Other *Mathematics and Its Applications* titles of interest:

J. Chaillou: *Hyperbolic Differential Polynomials and their Singular Perturbations.* 1979, 184 pp. ISBN 90-277-1032-5

S. Fucik: *Solvability of Nonlinear Equations and Boundary Value Problems.* 1981, 404 pp. ISBN 90-277-1077-5

V.I. Istratescu: *Fixed Point Theory. An Introduction.* 1981, 488 pp. *out of print,* ISBN 90-277-1224-7

F. Langouche, D. Roekaerts and E. Tirapegui: *Functional Integration and Semiclassical Expansions.* 1982, 328 pp. ISBN 90-277-1472-X

N.E. Hurt: *Geometric Quantization in Action. Applications of Harmonic Analysis in Quantum Statistical Mechanics and Quantum Field Theory.* 1982, 352 pp. ISBN 90-277-1426-6

F.H. Vasilescu: *Analytic Functional Calculus and Spectral Decompositions.* 1983, 392 pp. ISBN 90-277-1376-6

W. Kecs: *The Convolution Product and Some Applications.* 1983, 352 pp. ISBN 90-277-1409-6

C.P. Bruter, A. Aragnol and A. Lichnerowicz (eds.): *Bifurcation Theory, Mechanics and Physics. Mathematical Developments and Applications.* 1983, 400 pp. *out of print,* ISBN 90-277-1631-5

J. Aczel (ed.) : *Functional Equations: History, Applications and Theory.* 1984, 256 pp. *out of print,* ISBN 90-277-1706-0

P.E.T. Jorgensen and R.T. Moore: *Operator Commutation Relations.* 1984, 512 pp. ISBN 90-277-1710-9

D.S. Mitrinovic and J.D. Keckic: *The Cauchy Method of Residues. Theory and Applications.* 1984, 376 pp. ISBN 90-277-1623-4

R.A. Askey, T.H. Koornwinder and W. Schempp (eds.): *Special Functions: Group Theoretical Aspects and Applications.* 1984, 352 pp. ISBN 90-277-1822-9

R. Bellman and G. Adomian: *Partial Differential Equations. New Methods for their Treatment and Solution.* 1984, 308 pp. ISBN 90-277-1681-1

S. Rolewicz: *Metric Linear Spaces.* 1985, 472 pp. ISBN 90-277-1480-0

Y. Cherruault: *Mathematical Modelling in Biomedicine.* 1986, 276 pp. ISBN 90-277-2149-1

R.E. Bellman and R.S. Roth: *Methods in Approximation. Techniques for Mathematical Modelling.* 1986, 240 pp. ISBN 90-277-2188-2

R. Bellman and R. Vasudevan: *Wave Propagation. An Invariant Imbedding Approach.* 1986, 384 pp. ISBN 90-277-1766-4

A.G. Ramm: *Scattering by Obstacles.* 1986, 440 pp. ISBN 90-277-2103-3

Other *Mathematics and Its Applications* titles of interest:

C.W. Kilmister (ed.): *Disequilibrium and Self-Organisation.* 1986, 320 pp.
ISBN 90-277-2300-1

A.M. Krall: *Applied Analysis.* 1986, 576 pp.
ISBN 90-277-2328-1 (hb), ISBN 90-277-2342-7 (pb)

J.A. Dubinskij: *Sobolev Spaces of Infinite Order and Differential Equations.* 1986, 164 pp. ISBN 90-277-2147-5

H. Triebel: *Analysis and Mathematical Physics.* 1987, 484 pp.
ISBN 90-277-2077-0

B.A. Kupershmidt: *Elements of Superintegrable Systems. Basic Techniques and Results.* 1987, 206 pp. ISBN 90-277-2434-2

M. Gregus: *Third Order Linear Differential Equations.* 1987, 288 pp.
ISBN 90-277-2193-9

M.S. Birman and M.Z. Solomjak: *Spectral Theory of Self-Adjoint Operators in Hilbert Space.* 1987, 320 pp. ISBN 90-277-2179-3

V.I. Istratescu: *Inner Product Structures. Theory and Applications.* 1987, 912 pp.
ISBN 90-277-2182-3

R. Vich: *Z Transform Theory and Applications.* 1987, 260 pp.
ISBN 90-277-1917-9

N.V. Krylov: *Nonlinear Elliptic and Parabolic Equations of the Second -Order.* 1987, 480 pp. ISBN 90-277-2289-7

W.I. Fushchich and A.G. Nikitin: *Symmetries of Maxwell's Equations.* 1987, 228 pp. ISBN 90-277-2320-6

P.S. Bullen, D.S. Mitrinovic and P.M. Vasic (eds.): *Means and their Inequalities.* 1987, 480 pp. ISBN 90-277-2629-9

V.A. Marchenko: *Nonlinear Equations and Operator Algebras.* 1987, 176 pp.
ISBN 90-277-2654-X

Yu.L. Rodin: *The Riemann Boundary Problem on Riemann Surfaces.* 1988, 216 pp.
ISBN 90-277-2653-1

A. Cuyt (ed.): *Nonlinear Numerical Methods and Rational Approximation.* 1988, 480 pp. ISBN 90-277-2669-8

D. Przeworska-Rolewicz: *Algebraic Analysis.* 1988, 640 pp. ISBN 90-277-2443-1

V.S. Vladimirov, YU.N. Drozzinov and B.I. Zavialov: *Tauberian Theorems for Generalized Functions.* 1988, 312 pp. ISBN 90-277-2383-4

G. Morosanu: *Nonlinear Evolution Equations and Applications.* 1988, 352 pp.
ISBN 90-277-2486-5

Other *Mathematics and Its Applications* titles of interest:

A.F. Filippov: *Differential Equations with Discontinuous Righthand Sides.* 1988, 320 pp. ISBN 90-277-2699-X

A.T. Fomenko: *Integrability and Nonintegrability in Geometry and Mechanics.* 1988, 360 pp. ISBN 90-277-2818-6

G. Adomian: *Nonlinear Stochastic Systems Theory and Applications to Physics.* 1988, 244 pp. ISBN 90-277-2525-X

A. Tesar and Ludovt Fillo: *Transfer Matrix Method.* 1988, 260 pp. ISBN 90-277-2590-X

A. Kaneko: *Introduction to the Theory of Hyperfunctions.* 1989, 472 pp. ISBN 90-277-2837-2

D.S. Mitrinovic, J.E. Pecaric and V. Volenec: *Recent Advances in Geometric Inequalities.* 1989, 734 pp. ISBN 90-277-2565-9

A.W. Leung: *Systems of Nonlinear PDEs: Applications to Biology and Engineering.* 1989, 424 pp. ISBN 0-7923-0138-2

N.E. Hurt: *Phase Retrieval and Zero Crossings: Mathematical Methods in Image Reconstruction.* 1989, 320 pp. ISBN 0-7923-0210-9

V.I. Fabrikant: *Applications of Potential Theory in Mechanics. A Selection of New Results.* 1989, 484 pp. ISBN 0-7923-0173-0

R. Feistel and W. Ebeling: *Evolution of Complex Systems. Selforganization, Entropy and Development.* 1989, 248 pp. ISBN 90-277-2666-3

S.M. Ermakov, V.V. Nekrutkin and A.S. Sipin: *Random Processes for Classical Equations of Mathematical Physics.* 1989, 304 pp. ISBN 0-7923-0036-X

B.A. Plamenevskii: *Algebras of Pseudodifferential Operators.* 1989, 304 pp. ISBN 0-7923-0231-1

N. Bakhvalov and G. Panasenko: *Homogenisation: Averaging Processes in Periodic Media. Mathematical Problems in the Mechanics of Composite Materials.* 1989, 404 pp. ISBN 0-7923-0049-1

A.Ya. Helemskii: *The Homology of Banach and Topological Algebras.* 1989, 356 pp. ISBN 0-7923-0217-6

M. Toda: *Nonlinear Waves and Solitons.* 1989, 386 pp. ISBN 0-7923-0442-X

M.I. Rabinovich and D.I. Trubetskov: *Oscillations and Waves in Linear and Nonlinear Systems.* 1989, 600 pp. ISBN 0-7923-0445-4

A. Crumeyrolle: *Orthogonal and Symplectic Clifford Algebras. Spinor Structures.* 1990, 364 pp. ISBN 0-7923-0541-8

V. Goldshtein and Yu. Reshetnyak: *Quasiconformal Mappings and Sobolev Spaces.* 1990, 392 pp. ISBN 0-7923-0543-4

Other *Mathematics and Its Applications* titles of interest:

I.H. Dimovski: *Convolutional Calculus.* 1990, 208 pp. ISBN 0-7923-0623-6

Y.M. Svirezhev and V.P. Pasekov: *Fundamentals of Mathematical Evolutionary Genetics.* 1990, 384 pp. ISBN 90-277-2772-4

S. Levendorskii: *Asymptotic Distribution of Eigenvalues of Differential Operators.* 1991, 297 pp. ISBN 0-7923-0539-6

V.G. Makhankov: *Soliton Phenomenology.* 1990, 461 pp. ISBN 90-277-2830-5

I. Cioranescu: *Geometry of Banach Spaces, Duality Mappings and Nonlinear Problems.* 1990, 274 pp. ISBN 0-7923-0910-3

B.I. Sendov: *Hausdorff Approximation.* 1990, 384 pp. ISBN 0-7923-0901-4

A.B. Venkov: *Spectral Theory of Automorphic Functions and Its Applications.* 1991, 280 pp. ISBN 0-7923-0487-X

V.I. Arnold: *Singularities of Caustics and Wave Fronts.* 1990, 274 pp. ISBN 0-7923-1038-1

A.A. Pankov: *Bounded and Almost Periodic Solutions of Nonlinear Operator Differential Equations.* 1990, 232 pp. ISBN 0-7923-0585-X

A.S. Davydov: *Solitons in Molecular Systems. Second Edition.* 1991, 428 pp. ISBN 0-7923-1029-2

B.M. Levitan and I.S. Sargsjan: *Sturm-Liouville and Dirac Operators.* 1991, 362 pp. ISBN 0-7923-0992-8

V.I. Gorbachuk and M.L. Gorbachuk: *Boundary Value Problems for Operator Differential Equations.* 1991, 376 pp. ISBN 0-7923-0381-4

Y.S. Samoilenko: *Spectral Theory of Families of Self-Adjoint Operators.* 1991, 309 pp. ISBN 0-7923-0703-8

B.I. Golubov A.V. Efimov and V.A. Scvortsov: *Walsh Series and Transforms.* 1991, 382 pp. ISBN 0-7923-1100-0

V. Laksmikantham, V.M. Matrosov and S. Sivasundaram: *Vector Lyapunov Functions and Stability Analysis of Nonlinear Systems.* 1991, 250 pp. ISBN 0-7923-1152-3

F.A. Berezin and M.A. Shubin: *The Schrödinger Equation.* 1991, 556 pp. ISBN 0-7923-1218-X

D.S. Mitrinovic, J.E. Pecaric and A.M. Fink: *Inequalities Involving Functions and their Integrals and Derivatives.* 1991, 588 pp. ISBN 0-7923-1330-5

Julii A. Dubinskii: *Analytic Pseudo-Differential Operators and their Applications.* 1991, 252 pp. ISBN 0-7923-1296-1

V.I. Fabrikant: *Mixed Boundary Value Problems in Potential Theory and their Applications.* 1991, 452 pp. ISBN 0-7923-1157-4

Other *Mathematics and Its Applications* titles of interest:

A.M. Samoilenko: *Elements of the Mathematical Theory of Multi-Frequency Oscillations.* 1991, 314 pp. ISBN 0-7923-1438-7

Yu.L. Dalecky and S.V. Fomin: *Measures and Differential Equations in Infinite-Dimensional Space.* 1991, 338 pp. ISBN 0-7923-1517-0

W. Mlak: *Hilbert Space and Operator Theory.* 1991, 296 pp. ISBN 0-7923-1042-X

N.Ja. Vilenkin and A.U. Klimyk: *Representation of Lie Groups and Special Functions. Volume 1: Simplest Lie Groups, Special Functions, and Integral Transforms.* 1991, 608 pp. ISBN 0-7923-1466-2

N.Ja. Vilenkin and A.U. Klimyk: *Representation of Lie Groups and Special Functions. Volume 2: Class I Representations, Special Functions, and Integral Transforms.* 1992, 630 pp. ISBN 0-7923-1492-1

N.Ja. Vilenkin and A.U. Klimyk: *Representation of Lie Groups and Special Functions. Volume 3: Classical and Quantum Groups and Special Functions.* 1992, 650 pp. ISBN 0-7923-1493-X
(Set ISBN for Vols. 1, 2 and 3: 0-7923-1494-8)

K. Gopalsamy: *Stability and Oscillations in Delay Differential Equations of Population Dynamics.* 1992, 502 pp. ISBN 0-7923-1594-4

N.M. Korobov: *Exponential Sums and their Applications.* 1992, 210 pp. ISBN 0-7923-1647-9

Chuang-Gan Hu and Chung-Chun Yang: *Vector-Valued Functions and their Applications.* 1991, 172 pp. ISBN 0-7923-1605-3

Z. Szmydt and B. Ziemian: *The Mellin Transformation and Fuchsian Type Partial Differential Equations.* 1992, 224 pp. ISBN 0-7923-1683-5

L.I. Ronkin: *Functions of Completely Regular Growth.* 1992, 394 pp. ISBN 0-7923-1677-0

R. Delanghe, F. Sommen and V. Soucek: *Clifford Algebra and Spinor-valued Functions. A Function Theory of the Dirac Operator.* 1992, 486 pp. ISBN 0-7923-0229-X

A. Tempelman: *Ergodic Theorems for Group Actions.* 1992, 400 pp. ISBN 0-7923-1717-3

D. Bainov and P. Simenov: *Integral Inequalities and Applications.* 1992, 426 pp. ISBN 0-7923-1714-9

I. Imai: *Applied Hyperfunction Theory.* 1992, 460 pp. ISBN 0-7923-1507-3

Yu.I. Neimark and P.S. Landa: *Stochastic and Chaotic Oscillations.* 1992, 502 pp. ISBN 0-7923-1530-8

H.M. Srivastava and R.G. Buschman: *Theory and Applications of Convolution Integral Equations.* 1992, 240 pp. ISBN 0-7923-1891-9

Other *Mathematics and Its Applications* titles of interest:

A. van der Burgh and J. Simonis (eds.): *Topics in Engineering Mathematics.* 1992, 266 pp. ISBN 0-7923-2005-3

F. Neuman: *Global Properties of Linear Ordinary Differential Equations.* 1992, 320 pp. ISBN 0-7923-1269-4

A. Dvurecenskij: *Gleason's Theorem and its Applications.* 1992, 334 pp. ISBN 0-7923-1990-7

D.S. Mitrinovic, J.E. Pecaric and A.M. Fink: *Classical and New Inequalities in Analysis.* 1992, 740 pp. ISBN 0-7923-2064-6

H.M. Hapaev: *Averaging in Stability Theory.* 1992, 280 pp. ISBN 0-7923-1581-2

S. Gindinkin and L.R. Volevich: *The Method of Newton's Polyhedron in the Theory of PDE's.* 1992, 276 pp. ISBN 0-7923-2037-9

Yu.A. Mitropolsky, A.M. Samoilenko and D.I. Martinyuk: *Systems of Evolution Equations with Periodic and Quasiperiodic Coefficients.* 1992, 280 pp. ISBN 0-7923-2054-9

I.T. Kiguradze and T.A. Chanturia: *Asymptotic Properties of Solutions of Nonautonomous Ordinary Differential Equations.* 1992, 332 pp. ISBN 0-7923-2059-X

V.L. Kocic and G. Ladas: *Global Behavior of Nonlinear Difference Equations of Higher Order with Applications.* 1993, 228 pp. ISBN 0-7923-2286-X

S. Levendorskii: *Degenerate Elliptic Equations.* 1993, 445 pp. ISBN 0-7923-2305-X

D. Mitrinovic and J.D. Kečkić: *The Cauchy Method of Residues, Volume 2.* Theory and Applications. 1993, 202 pp. ISBN 0-7923-2311-8

R.P. Agarwal and P.J.Y Wong: *Error Inequalities in Polynomial Interpolation and Their Applications.* 1993, 376 pp. ISBN 0-7923-2337-8

A.G. Butkovskiy and L.M. Pustyl'nikov (eds.): *Characteristics of Distributed-Parameter Systems.* 1993, 386 pp. ISBN 0-7923-2499-4

B. Sternin and V. Shatalov: *Differential Equations on Complex Manifolds.* 1994, 504 pp. ISBN 0-7923-2710-1

S.B. Yakubovich and Y.F. Luchko: *The Hypergeometric Approach to Integral Transforms and Convolutions.* 1994, 324 pp. ISBN 0-7923-2856-6

C. Gu, X. Ding and C.-C. Yang: *Partial Differential Equations in China.* 1994, 181 pp. ISBN 0-7923-2857-4

V.G. Kravchenko and G.S. Litvinchuk: *Introduction to the Theory of Singular Integral Operators with Shift.* 1994, 288 pp. ISBN 0-7923-2864-7

A. Cuyt (ed.): *Nonlinear Numerical Methods and Rational Approximation II.* 1994, 446 pp. ISBN 0-7923-2967-8

Other *Mathematics and Its Applications* titles of interest:

A.J. Jerri: *Linear Difference Equations with Discrete Transform Methods*. 1996, 462 pp. ISBN 0-7923-3940-1

I. Novikov and E. Semenov: *Haar Series and Linear Operators*. 1996, 240 pp. ISBN 0-7923-4006-X

L. Zhizhiashvili: *Trigonometric Fourier Series and Their Conjugates*. 1996, 312 pp. ISBN 0-7923-4088-4

R.G. Buschman: *Integral Transformation, Operational Calculus, and Generalized Functions*. 1996, 246 pp. ISBN 0-7923-4183-X

V. Lakshmikantham, S. Sivasundaram and B. Kaymakcalan: *Dynamic Systems on Measure Chains*. 1996, 296 pp. ISBN 0-7923-4116-3

GPSR Compliance

The European Union's (EU) General Product Safety Regulation (GPSR) is a set of rules that requires consumer products to be safe and our obligations to ensure this.

If you have any concerns about our products, you can contact us on ProductSafety@springernature.com

In case Publisher is established outside the EU, the EU authorized representative is:

Springer Nature Customer Service Center GmbH
Europaplatz 3
69115 Heidelberg, Germany

Batch number: 10620084

Printed by Printforce, the Netherlands